CONSERVATION TILLAGE SYSTEMS AND WATER PRODUCTIVITY IMPLICATIONS FOR SMALLHOLDER FARMERS IN SEMI-ARID ETHIOPIA

Conservation Tillage Systems and Water Productivity Implications for Smallholder Farmers in Semi-arid Ethiopia

DISSERTATION

Submitted in fulfilment of the requirements of
the Board for Doctorates of Delft University of Technology and of
the Academic Board of the UNESCO-IHE Institute for Water Education for the
Degree of DOCTOR
to be defended in public
on Tuesday, 20 February 2007 at 10:00 hours
in Delft, the Netherlands

by

Melesse Temesgen Leye
Master of Science, Ethiopian Agricultural Research Institute

born in Gojjam, Ethiopia

This dissertation has been approved by the promotor
Prof.dr.ir. H.H.G. Savenije   TU Delft/UNESCO-IHE Delft, The Netherlands

Taylor & Francis is an imprint of the Taylor & Francis Group, an informa business
© 2007, Melesse Temesgen

Published by:
Taylor & Francis/Balkema
PO Box 447, 2300 AK Leiden, The Netherlands
e-mail: Pub.NL@tandf.co.uk
www.balkema.nl, www.taylorandfrancis.co.uk, www.crcpress.com

ISBN10 0-415-43946-9 (Taylor & Francis Group)
ISBN13 978-0-415-43946-6 (Taylor & Francis Group)

Synopsis

The traditional tillage implement, the *Maresha* plow, and the tillage systems that require repeated and cross plowing have caused poor rainfall partitioning and hence low water productivity in Ethiopia. Considering the limitations to a wider application of irrigation schemes among the resource poor farmers, it is critical that the available rainwater is managed properly by increasing infiltration and water holding capacity of the soil and by minimizing evaporation losses especially during the dry periods. Conservation tillage with no-till has been used by farmers in America (south and North) and Australia to alleviate these problems. However, since no-till could not be easily adopted by smallholder farmers in Africa, due to socio-economic and environmental problems, locally adapted conservation tillage systems have been introduced in several parts of the continent in order to improve labor, soil and water productivity. Direct application of these techniques to Ethiopia was again constrained by the fact that the conservation tillage implements developed for farmers in other African countries do not fit on the frames of the traditional tillage implement used in Ethiopia. Furthermore, the unique crop in Ethiopia, *tef (Eragrostis tef* (Zucc) Trotter), has to be broadcast, which makes it difficult to apply conservation tillage systems developed for row planted crops.

This thesis reports on research carried out to evaluate alternative conservation tillage systems suitable for the smallholder farmers in semi-arid Ethiopia. Surveys were carried out in order to study the traditional tillage systems in two selected semi-arid areas and to identify reasons for repeated plowing. Implements that were developed as modifications or attachments to the traditional tillage implement, the *Maresha* Plow, have been tested in the field in order to evaluate their suitability for undertaking conservation tillage. Strip tillage systems for maize production that involve opening of furrows with the *Maresha* Plow followed by planting along the cultivated lines with subsoiling (STS) or without subsoiling (ST) were compared with the traditional tillage system (CONV). Improved tillage systems for *tef* production that involve plowing once with the *Maresha* Modified Plow and the use of the Sweep at planting with subsoiling (ITS) or without subsoiling (IT) have been compared with CONV. Assessment of the different tillage systems were made on water productivity and profitability. Daily rainfall, surface runoff and soil moisture were directly measured while a physically based model and a conceptual threshold model were used to estimate the water balance components.

It was realized that repeated plowing is caused by the V-shaped furrow created by the *Maresha* plow, which leaves unplowed strip of land between adjacent passes. Farmers are forced to carry out cross plowing to disturb the unplowed land. Cross plowing on steep slopes can cause high surface runoff when the furrows are laid along the slope. According to farmers, the main purpose of tillage is to conserve moisture and to control weeds. Dry spells occurring between rainfall events create surface crusts and allow emergence of weeds thus forcing farmers to plow frequently. Soil warming is also perceived as a purpose of tillage but this needs further investigation.

The conservation tillage implements that were developed as modifications to the traditional tillage implement, the *Maresha* Plow, were found to be suited to the respective operations they were developed for. The Subsoiler disrupted the plow pan while the Tie-ridger made furrows of larger cross sectional areas than those made by the *Maresha* Plow and the inverted Broad Bed Maker (BBM), with lower pulling and lifting forces. Planting maize with the Row Planter resulted in twice as much seedling emergence as manual placement of seeds leading to increased grain yields in addition to saving labor and time. The *Maresha* Modified Plow made U-shaped furrows and controlled weeds better than the *Maresha* Plow.

Among the conservation tillage systems tested on *tef*, ITS resulted in the least surface runoff (Qs=23 mm-season^{-1}), the highest crop transpiration (T=53 mm-season^{-1}), the highest grain yields (Y=1180 kg-ha^{-1}) and the highest water productivity using total evaporation (W_{PET}=0.42 kg-m^{-3}) followed by CONV (Qs=34 mm-season^{-1}, T=49 mm-season^{-1}, Y=1070 kg-ha^{-1}, W_{PET}=0.39 kg-m^{-3}) and MT (Qs=48 mm-season^{-1}, T=32 mm-season^{-1}, Y=890 kg-ha^{-1}, W_{PET}=0.32 kg-m^{-3}). Among the conservation tillage treatments tested on maize, STS resulted in the least surface runoff (Qs=17 mm-season^{-1}), the highest transpiration (T=196 mm-season^{-1}), the highest grain yields (Y=2130 kg-ha^{-1}) and the highest water productivity using total evaporation (W_{PET}=0.67 kg-m^{-3}) followed by ST (Qs=25 mm-season^{-1}, T=178 mm-season^{-1}, Y=1840 kg-ha^{-1}, W_{PET}=0.60 kg-m^{-3}) and CONV (Qs=40 mm-season^{-1}, T=158 mm-season^{-1}, Y=1720 kg-ha^{-1}, W_{PET}=0.58 kg-m^{-3}). However, when the time between the last tillage operation and planting of maize was more than 26 days the reverse occurred. There was no statistically significant change in soil physical and chemical properties after three years of experimenting with different tillage systems.

A simple conceptual model simulated soil moisture in the root zone better than a physically based model that employed Richards equations because preferential flows are important in the semi-arid tropics whereas the physically based model did not consider such flows.

The experiments have shown that it is indeed possible to introduce new insights and new technology into traditional farming systems, provided these innovations increase yields, reduce labor and are affordable. It is concluded that the locally adapted conservation tillage technologies can help the resource poor smallholder farmers in semi arid areas of Ethiopia achieve food security by positively altering rainfall partitioning and by improving water productivity. The improved implements and the conservation tillage system tested on *tef*, ITS, can be popularized among farmers while additional trials are required to verify the performance of STS, paying particular attention to the time of subsoiling.

Table of Contents

Abbreviations and Acronyms

ACT	African Conservation Tillage Network
BBM	Broad Bed and furrow Maker
BD	Bulk Density (gm-cm^{-3})
CADU	Chilalo Agricultural Development Unit
CONV	Conventional Tillage
CSA	Central Statistics Authority
CTIC	Conservation Tillage Information Center
EARI	Ethiopian Agricultural Research Institute
FAO	Food and Agriculture Organization
GDP	Gross domestic product (Birr)
IAR	Institute of Agricultural Research
IGAD	Intergovernmental Authority on Development
IIRR	International Institute for Rural Reconstruction
IT	Improved tillage
ITS	Improved tillage with subsoiling
MMP	*Maresha* Modified Plow
MT	Minimum Tillage
SCH	Sealing, Crusting and Hardsetting
SG2000	Sasakawa Global 2000
SOC	Soil Organic Carbon (%)
ST	Strip tillage
STD	Standard deviation
STS	Strip tillage with subsoiling
TDR	Time-Domain Reflectometer
TN	Total Nitrogen (%)
DTP	Days between last tillage and planting
UNDP	United Nations Development Program
WOTRO	Dutch Foundation for the Advancement of Tropical Research

Notations and Symbols

Notation	Description	Dimension
A	The leaf area	L^2
C_C	Correction factor to take account of leaf overlaps	-
C_p	Specific heat of air at constant pressure	$L^2T^{-2\circ}C^{-1}$
D	Interception threshold	LT^{-1}
E	Total evaporation	LT^{-1}
E_s	Evaporation from the soil	LT^{-1}
I	Interception	LT^{-1}
I_{LA}	The leaf area index	L^2L^{-2}
K_C	Crop factor determining the proportion of crop transpiration to open water evaporation.	-
K_R	A parameter that takes account of the share of deep percolation from storage in the root zone	-
L	Leaf length	L
N	The number of leaves in each plant.	-
P	Precipitation	LT^{-1}
P_{NET}	Net rainfall obtained by subtracting a runoff threshold	LT^{-1}
P_0	Plant population	L^{-2}
Q_s	Surface runoff	LT^{-1}
R	Deep percolation	LT^{-1}
R_{is}	Incoming short-wave solar radiation	MT^{-3}
R_n	Net radiation available for evaporation	MT^{-3}
S	Soil water storage in the root zone	L
S_{FC}	Stored soil water at field capacity	L
S_W	Stored soil water at wilting point	L
T	Transpiration by the plant	LT^{-1}
T_0	Plant transpiration when there is no limitation in soil moisture	LT^{-1}
W_M	The maximum width of the leaf	L
W_{PT}	Water productivity using transpiration	ML^{-3}
W_{PET}	Water productivity using total evaporation	ML^{-3}
W_{PP}	Water productivity using rainfall	ML^{-3}
X_i	The vector of independent variables	-
Y	Grain yield	ML^{-2}
Y_i	The intensity of tillage calculated as the ratio of each tillage frequency to the maximum tillage frequency	-
Y_i^*	An underlying latent variable that indexes the intensity of tillage	-
d	Displacement height	L
e_a	Actual vapor pressure	$ML^{-1}T^{-2}$
e_s	Vapor pressure at saturation	$ML^{-1}T^{-2}$
f_d	Displacement height factor	-

$f(\Psi(z))$	A factor used to reduce T if the soil water tension in the root zone drops below a critical value, Ψ_c.	-
g_l	Stomatal conductance	LT^{-1}
$g_{max,}$	Maximum canopy conductance	LT^{-1}
g_{ris}	ris representing half light saturation	MT^{-3}
g_{vpd}	vpd corresponding to 50% reduction in stomatal conductance	$MT^{-2}L^{-1}$
k	Von Karman's constant	-
k_w	Unsaturated hydraulic conductivity	LT^{-1}
p_1	Water uptake reduction coefficient 1	$L^{-3}T^{-1}$
p_2	Water uptake reduction coefficient 2	$ML^{-2}T^{-1}$
r_a	Aerodynamic resistance.	-
ris	Global radiation intensity	MT^{-3}
r_s	Surface resistance	TL^{-1}
r_{sc}	Plant surface resistance for the canopy	TL^{-1}
r_{ss}	Soil surface resistance	L
$r_{\psi1}, r_{\psi2}$	Empirical coefficients for surface resistance	TL^{-1}
u	Wind speed	LT^{-1}
vpd	Vapor pressure deficit	$MT^{-2}L^{-1}$
z	Soil depth	L
z_o	Roughness length.	L
z_{ref}	Reference height	L
Δ	Slope of saturated vapour pressure versus temperature curve	-
Ψ	Soil water tension	L
Ψ_c	Critical value of soil water tension for calculating T	L
Ψ_s	Water tension at the soil surface.	L
α	Coefficient for leaf area	-
β	A vector of parameters to be estimated	-
ε_i	An error term.	-
γ	Psychrometer constant	$ML^{-1}T^{-2}$
ρ_a	Air density	ML^{-3}
ρ_w	Density of water	ML^{-3}
λ	Latent heat of vaporization	L^2T^{-2}
$1-p$	The fraction of soil water available to the crop ($S_{FC}-S_W$) below which transpiration is limited by moisture stress	-

Chapter 1

INTRODUCTION

1.1 Overview

Land productivity in many parts of sub-Saharan Africa is declining (Middleton and Thomas, 1997). Crop yields from staple food crops such as maize, millet and sorghum remain in the order of 1 t grain ha^{-1} in smallholder rain fed farms (Rockström and Jonsson, 1999). There is an urgent need for the introduction of sustainable soil management practices in order to reverse the food crises in sub-Saharan Africa.

In the semi-arid regions of Africa, short intense storms coupled with prolonged dry spells make crop production difficult. Intensive rainfall causes a high proportion of surface runoff that also carries away the top fertile soil. Due to high temperatures, soil evaporation can reach 30-50% of the total rainfall leaving only 10-30% for crop transpiration (Figure 1.1). Poor rainfall partitioning leads to low water productivity. Considering the limitations to a wider application of irrigation schemes among the resource poor farmers, it is critical that the

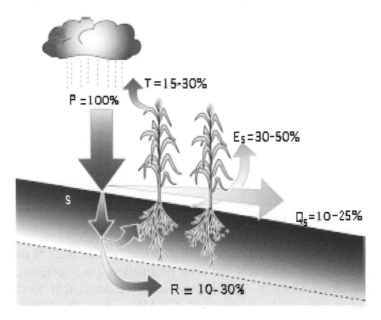

Figure.1.1 General overview of rainfall partitioning in farmers' fields in semi-arid savannah agro ecosystems in sub-Saharan Africa. P = seasonal rainfall, E_S = soil evaporation and interception, S = soil moisture, T = plant transpiration, Q_S = surface runoff and R = deep percolation. (Adapted from: Rockström *et al.*, 2001)

available rainwater is managed properly by increasing infiltration and water holding capacity of the soil and by minimizing evaporation losses especially during the dry periods.

Ethiopia is located in the Horn of Africa between 3^0 and 18^0 North latitude and 33^0 and 48^0 East longitude (Figure 1.2). Its population is currently estimated to be more than 70 million. Agriculture is the mainstay of the country's economy with 60% of GDP coming from the sector. It is a means of livelihood for about 85% of the total population. The main power sources in agriculture have been human and animals. According to surveys conducted earlier (Pathak, 1987), over 90% of the total agricultural produce comes from 5.5 million farmers employing 5 million oxen and cultivating 95% of the land under plow.

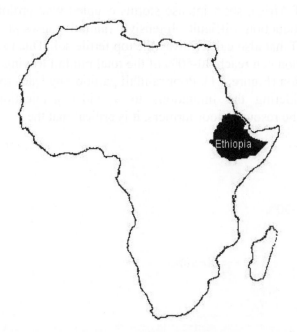

Figure 1.2. The location of Ethiopia in Africa.

The semi-arid areas in Ethiopia (Figure 1.3) cover 301,500 km^2, which is 27 % of the country. The semi-arid areas represent the crop production zone suffering from a serious moisture stress (Engida, 2000). It is in these areas that food insecurity and famine has always been reported (IGAD and FAO, 1995). Shortage of rainfall is normally reported as the cause of famine in Ethiopia. However, the total rainfall in the semi-arid areas can be as high as 700 mm-yr^{-1} (Figure 3.1). One could ask why so much annual rainfall wouldn't be sufficient to grow crops. The underlying reason for the inability of farmers to feed themselves in these areas is the high proportion of losses mainly as a result of surface runoff and soil evaporation as demonstrated in Figure 1.1. In addition to the environmental factors, the causes of poor rainfall partitioning are believed to be poor soil management with traditional tillage systems (Rockström *et al.*, 2001).

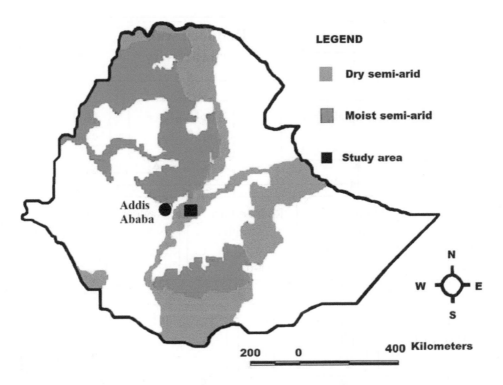

Figure 1.3. Semi arid areas in Ethiopia. (Source: IGAD and FAO, 1995). Areas with length of growing period in the range of 60 to 119 days are classified as dry semi-arid while areas with a length of growing period of 120-179 days are classified as moist semi-arid.

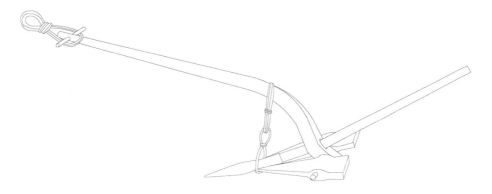

Figure 1.4. The traditional tillage implement, the *Maresha* plow.

1.2 The 'Cause and Effect Tree' of Traditional Tillage and Low Water Productivity

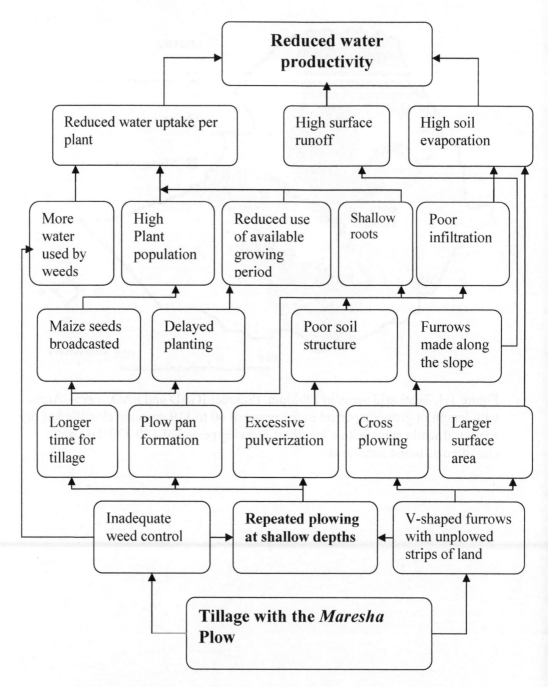

Figure 1.5. Cause and effect tree showing how traditional tillage systems with the *Maresha* plow contribute to low water productivity in the dry semi arid regions of Ethiopia.

The traditional tillage implement in Ethiopia, the *Maresha* plow (Figure 1.4), and the related tillage system that requires repeated plowing have caused reduced water productivity through a number of ways (Figure 1.5).

Traditional tillage with the *Maresha* plow requires repeated plowing with any two consecutive tillage operations carried out perpendicular to each other. Such a practice of cross plowing is necessary because of the V-shaped furrows created by the *Maresha* plow (Figure 3.2). Since a certain area of land is left undisturbed in the first pass, farmers have to do a second tillage with furrows intersecting each other in order to access the unplowed part. Otherwise, the plow will slip into the previously made furrows thereby missing the unplowed strips. In moderate to steep slopes, one of any two consecutive tillage operations will be oriented along or nearly along the slope thus encouraging runoff. With the very hilly topography of much of Ethiopia such tillage systems have caused large losses of soil and water through runoff.

The V-shaped furrows also result in higher relative surface area exposure leading to increased loss of moisture through evaporation. Rough surfaces resulting from primary tillage operations that enhance higher rates of gas exchanges were also identified to be causes of increased CO_2 emission (Reicosky, 2001) thus resulting in high losses of organic carbon.

As it can be seen in Figure 1.5, because of incomplete plowing by *Maresha*, farmers have to do repeated tillage in order to produce a fine seedbed especially for *tef*. As a result, the soil is excessively pulverized thus resulting in poor structure (crust formation, compaction, etc.). Moreover, the *Maresha* plow cuts the vegetative parts of grass weeds which can result in more propagation and multiplication. Hence, the inefficiency of *Maresha* in controlling grass weeds forces farmers to do repeated tillage. If weeds are not properly controlled, crop water uptake is reduced due to competitions. As a result of repeated tillage at shallow depth plow pans may form (Chapter 3) which hinder water infiltration (Whiteman, 1979) and root growth (Willcocks, 1984; Rowland, 1993).

During cross plowing, farmers are forced to run over the already tilled soil in an attempt to access the unplowed part. Consequently, they spend almost 50% of the time passing across already plowed furrows, which is even more during the third tillage that is aimed at reaching spots of unplowed land left after the second plowing. Such a repeated action also imparts a high amount of energy on some of the soil particles leading to localized excessive pulverization. Excessive pulverization damages the soil structure resulting in poor infiltration and shallow roots. Shallow rooted plants use smaller amounts of water for transpiration, which means reduced water productivity. On the other hand, poor soil structure causes lower infiltration rates (Rockström and Valentin, 1997; Hoogmoed, 1999) resulting in higher proportion of water lost through evaporation and surface runoff (Figure 1.5).

The other effect of repeated tillage with cross plowing is the longer time required for seedbed preparations, which delays planting (Pathak, 1987) making it difficult for the crop to fully utilize the available growing period. Moreover, farmers broadcast seeds of maize, instead of row planting, because of shortage of time resulting in high plant population and hence reduced water uptake per plant. Lower maize yields due to broadcasting of maize were reported by several investigators (Rowland, 1993). Figure 4.8 in chapter 4 shows poorly managed fields with maize seeds broadcasted and no fertilizer applied (right) while the one on the left is row planted maize with localized fertilizer application using animal drawn row seeder.

1.3 The Solution

Past research on conservation tillage in Ethiopia has concentrated mainly on reducing the number of tillage operations required for seedbed preparation. However, mere reduction of tillage frequency while using the same implement, the *Maresha* plow, and the same tillage system can compromise grain yields thus making the practice unacceptable. Farmers have probably experimented with different tillage frequencies over centuries and have chosen the type of tillage frequency that they are using at the moment. Reducing tillage frequency would have been farmers' preference considering the limitations they face in terms of time, labor and traction requirement. Therefore, research has to come up with a different tillage system and/or different implements that can reduce tillage frequency without compromising yields.

In semi arid regions, the key to increased local crop production is maximizing infiltration at the expense of surface runoff. Moreover, techniques that lead to reduced soil evaporation during dry spells increase the amount of water available to crops. One way of achieving such objectives could be the introduction of conservation tillage practices using appropriate equipment (Ahenkorah, *et al.,* 1995; Chen *et al.,* 1998; Steiner, 1998; Tiscareno *et al.,* 1999; Biamah and Rockström, 2000; Freitas, 2000). According to Rockström *et al* (2001), conservation tillage is any tillage system that conserves water and soil while saving labor and traction needs. The attempts made to adapt the definitions of conservation tillage to an African context have been influenced mainly by socio economic and environmental constraints to maintaining the required soil cover due to competition for crop residue by livestock and the inherently low bio-mass production in low rainfall areas. The benefits of no-tillage are realized with soil cover responsible for regulating soil temperature which also reduces evaporation losses, reducing the impact of rainfall thereby reducing compaction and soil erosion and reducing runoff by acting as barriers to water movement (Chapter 2). Consequently, most of the research and extension activities have focused on conservation tillage systems that involve ripping along planting lines and subsoiling to break plow pans created by the repeated action of the plow using improved implements such as the Magoye Ripper and the Palabana Subsoiler (Hoogmoed *et al.,* 2003).

1.4   Problems that are peculiar to Ethiopia

In many sub-Saharan African countries, maize, which can and often is planted in lines, is the staple food. Since maize is well suited for ripping based tillage in which only planting lines are cultivated while leaving the area in between undisturbed, such a tillage system has been successfully introduced (Biamah *et al.*, 1993; Steiner, 1998; Rockström *et al.*, 2001). However, there has been very little effort to develop conservation tillage system for broadcasted crops. In Ethiopia, the majority of smallholder farmers grow a small seeded cereal called *tef (Eragrostis tef* (Zucc) Trotter). *Tef* cannot be planted in rows unlike maize and wheat because the plant has a very small size that makes it unable to fully utilize spaces left between rows. Hence, there is a need to develop conservation tillage systems that are suitable for *tef* production.

Conservation tillage systems developed in other African countries use implements that were developed as modifications or attachments to the steel moldboard plows. In Ethiopia, animal traction for tillage is associated with a traditional plow known as *Maresha*. The *Maresha* plow has been used by the highlanders for thousands of years (Goe, 1987). It is very simple, light in weight, cheap, and is locally made. However, the conservation tillage implements developed for farmers in other African countries do not fit on the frames of the *Maresha* Plow. On the other hand, a number of improved tillage implements that can be used for conservation tillage have been developed as modifications or attachments to the *Maresha* Plow (Temesgen, 2000). This study was undertaken to evaluate different types of conservation tillage systems using the *Maresha* modified implements both for row planted and broadcasted crops.

1.5   Objectives

The general objective of this study is to increase the understanding of the traditional and improved tillage systems and implements and to evaluate alternative conservation tillage systems, in terms of their effects on yields, water productivity, labor needs and traction requirements for the smallholder farmers in semi-arid areas of Ethiopia.

The specific objectives of the study are:

1. To study when, how many times and why farmers undertake tillage for maize and *tef* production.
2. To study the existence and the nature of plow pans under the *Maresha* cultivation system.
3. To test the field performance of the improved tillage implements including: the Subsoiler, the Sweep, the Tie-Ridger, the Row planter and the *Maresha* Modified Plow.
4. To study the implications on surface runoff and water productivity of strip tillage systems for maize production.
5. To study the implications on surface runoff and water productivity of improved tillage systems for *tef* production.

6. To study the short term effects of the proposed conservation tillage systems on physical and chemical properties of soils.

1.6   Outline of the thesis

This thesis is structured into seven chapters. Chapter 1 introduces the problem that led to the study and the structure of the thesis.

In Chapter 2, a research review on conservation tillage (CT) is presented showing the historical developments and why one form of recipe can not be applied everywhere leading to the hypotheses of the study.

In Chapter 3, the traditional tillage systems as studied in two selected sites in a semi arid region of Ethiopia are presented. The studies reveal the rationale behind repeated tillage as perceived by farmers. The timing and purposes of each stage of tillage operation that are identified through the study can be useful in developing suitable conservation tillage systems that can maximize water productivity in semi-arid regions.

Chapter 4 presents the conservation tillage implements that were developed as modifications of the traditional *Maresha* Plow. Results of field testing and evaluation of each tillage implement in terms of draft power requirement, disruption of the plow pan for improved infiltration, weed control, reduction of tillage frequency, improving seedling emergence and achieving timeliness of operation are presented.

In Chapter 5, water productivity of strip tillage systems for maize production that were tested on farmers' fields is presented. A physically based CoupModel and a conceptual threshold model were used to estimate the water balance components as affected by the different tillage systems.

Chapter 6 presents the results of a three-year on-farm trial carried out to compare different types of tillage systems developed for broadcasted crops like *tef*. Assessments were made on water productivity and profitability. A conceptual threshold model was used to estimate water balance components in different tillage systems.

Chapter 7 gives conclusions and recommendations based on the findings of the study. Research questions that need to be addressed with a follow-up trial are presented.

Chapter 2

RESEARCH BACKGROUND AND HYPOTHESES

In Chapter 1, problems related to conventional tillage systems with the traditional tillage implement, the *Maresha* Plow, were presented. In this chapter, we will look at the more general problems of conventional tillage systems and review of research on conservation tillage in Ethiopia, in Africa and in the world.

2.1 Definitions of Conservation Tillage

Conservation tillage is defined by the Conservation Tillage Information Center (CTIC) as any tillage and planting system that covers 30 percent or more of the soil surface with crop residue, after planting, to reduce soil erosion by water. According to the European Conservation Agriculture Federation (ECAF), conservation agriculture refers to several practices which permit the management of soil for agrarian uses, altering its composition, structure and natural biodiversity as little as possible and protecting it from erosion and degradation (ECAF, 1999). Synonymous to conservation tillage are conservation farming and conservation agriculture. The Food and Agriculture Organization (FAO) uses the name Conservation Agriculture (CA) and defines CA as the simultaneous application of minimum soil disturbance, soil cover and crop rotation (Benites and Ashburner, 2001). The main justification for the change in the name to conservation agriculture is the inclusion of other issues such as crop rotation and soil cover while no tillage is advocated. A recent definition of CA by Dumanski *et al* (2006) states that the principles of CA include maintaining permanent soil cover, promoting a healthy, living soil, promoting balanced application and precision placement of fertilizers, pesticides, and other crop inputs, promoting legume fallows, composting, and organic soil amendments, and promoting agro-forestry to enhance on-farm biodiversity and alternate sources of income.

In Africa, the term conservation tillage has been used in a more flexible way to refer to any tillage system which conserves or reduces soil, water and nutrient loss or which reduces draft power requirements for crop production (Steiner, 1998). According to Rockström *et al* (2001), conservation tillage is any tillage system that conserves water and soil while saving labour and traction needs. The attempts made to adapt the definitions of conservation tillage to an African context have been influenced mainly by lack of soil cover as a result of competition for crop residue by livestock and the inherently low bio-mass production due to shortage of rainfall. In this thesis, emphasis has been given to conservation tillage systems that address the issue of water productivity by positively altering rainfall partitioning in the semi arid areas of Ethiopia.

2.2   The need for conservation tillage

2.2.1   On-farm water balance and water productivity

In this thesis, water productivity is considered equivalent to water use efficiency, which can be defined in many ways, but in general terms it refers to the amount of crop produced per unit of water expressed in $kg\text{-}m^{-3}$ where the yield is expressed in $kg\text{-}ha^{-1}$ and the water used expressed in $m^3\text{-}ha^{-1}$.

Rainfall partitioning (Figure 1.1) significantly affects the water balance and availability of water to crops (Rockström and Valentin, 1997). Tillage affects the two partitioning points in the water balance. The first partitioning point is where rainfall is partitioned at the soil surface into interception, infiltration, and surface runoff while the second partitioning point is where soil moisture is partitioned between crop water uptake, soil evaporation and drainage. Repeated conventional tillage damages the soil structure through excessive pulverization and mineralization leading to reduction in soil organic matter content and aggregate stability. This results in soil compaction over the plowed layer, surface crust and plow pan formation that reduce infiltration thus affecting the first partitioning point.

In the second partitioning point conventional tillage reduces water uptake by plants because root growth is restricted over and below the plowed layer. Moreover, the water holding capacity of the soil may be reduced through loss of organic matter and soil compaction, which results in less water available for useful transpiration by the crop. Conservation tillage is aimed at altering the rainfall partitioning such that more infiltration at the expense of surface runoff, in the first partitioning point, and more root water uptake thus more useful transpiration at the expense of soil evaporation, in the second partitioning point, are achieved (Rockström and Valentin, 1997).

2.2.2   Combating land degradation

A number of investigations have been carried out to observe the effect of tillage treatments on soil erosion and have concluded that intensive tillage exposes the soil to more erosion (Biamah and Rockström, 2000; Hoogmoed, 1999; Benites and Ashburner, 2001; Nitzsche *et al.,* 2001).

The effects of tillage can be seen as direct, in which the soil is made ready for transportation by water and wind through loosening, and as indirect, in which the soil is degraded in the form of crusting and surface sealing resulting in less infiltration causing runoff (Benites and Ashburner, 2001). The detachment of soil particles by the impact of rain drops is considerable since the sealing and crusting processes are caused by the instability of the soil aggregates at the surface. Soil degradation is generally seen as a result of erosion processes, but the underlying phenomena may be the sealing and crusting behavior of the soil (Hoogmoed, 1999) that reduces infiltration leading to high surface runoff and soil erosion (Rockström and Valentin, 1997; Hoogmoed, 1999).

The continuous removal of crop residues, coupled with minimal use of farmyard manure results in the mining of nutrients, organic matter depletion, and weakening of the soil structure. These processes lead to increased runoff and erosion losses that are strongly linked to loss of topsoil. This situation makes it even more difficult for any extra crop residue to be retained on cropland for soil and water conservation (Okwach and Simiyu, 1999).

Land degradation problems in the tropical climates are generally higher than those in temperate regions because organic matter reduction caused by intensive tillage is very fast in the former (Derpsch and Moriya, 1998). It is estimated that reductions in organic matter content to values below 1% and sometimes as low as 0.2% can be reached in only one or two decades of intensive soil preparation (Derpsch and Moriya, 1998; Jaiyeoba, 2003). Conservation tillage aims at combating soil degradation by reducing soil erosion and improving soil quality (Benites and Ashburner, 2001; Nitzsche *et al.,* 2001)

2.2.3 Increased crop production on a sustainable basis
Several researchers have reported increased yields from conservation tillage (e.g. Scopel, *et al.,* 2001; Diaz-Zorita *et al.,* 2002). The major reasons for the increase in yields were better moisture availability, early planting, improved soil fertility, better root growth and aeration. In addition, since conservation tillage reverses the process of land degradation by improving or maintaining soil quality, sustainable improvement in crop production is the main focus.

2.3   Requirements of conservation tillage
The full benefits of conservation tillage can be realized through minimum soil disturbance, soil cover and crop rotation. Although recent definitions (Dumanski *et al.,* 2006) have expanded the components of conservation agriculture, the following are considered to be the pillars of the ideal conservation tillage system.

2.3.1 Minimum soil disturbance
No till is the most preferred system provided conditions for its proper implementation are met. Intensive tillage exposes the soil to the various environmental effects leading to loss of soil organic carbon through oxidation, loss of soil moisture through evaporation and loss of soil and water through surface runoff. Therefore, any conservation tillage system should involve reduction of tillage in one way or another.

2.3.2 Soil cover
Soil cover can be made by growing cover crops or by leaving crop residues in the field. Cover crops are crops grown before or during the vegetative period of the main crop with the objective of improving soil organic matter and protecting the soil from adverse effects of the environment such as erosion. Moreover, cover crops add organic matter, improve soil structure and tilth, fix atmospheric nitrogen, recycle unused soil nitrogen, increase soil productivity and suppress weeds (Wilson *et al.,* 1982; Tsai *et al.* 1989). A cover crop provides vegetative cover during periods when a crop is not present to deflect the force of falling raindrops, which otherwise would detach soil particles and make them prone to

erosion. It also slows down the rate of runoff, thus improving moisture infiltration into the soil (Benites and Ashburner, 2001; Martinez-Raya *et al.*, 2001; Fuentes *et al.*, 2003). Plant residues reduce water runoff and wind erosion by preserving surface soil structure (Addiscott and Dexter, 1994; Papendick and McCool, 1994).

2.3.3 Enhanced soil organic matter

Organic matter has a strong positive effect on infiltration of water into soils. This effect is due mainly to a decrease in bulk density, and improvements in aggregation and structure. In soils that have been cultivated for a long time, the organic matter content and fertility is very low, the soil structure is poor and compaction is high. Such initial conditions make it difficult to directly go into zero tillage as this necessitates treatments with soil organic matter enhancing activities such as green manuring (Elwell *et al.*, 2000). Green manure crops are crops grown to be directly incorporated for the purpose of enriching agricultural soil. Green manuring provides highly effective weed control, increased nutrient availability in the following year, and improved soil organic matter (Monjardino *et al.*, 2000). The benefits of green manuring on crop yield are most apparent during dry periods, particularly in rain fed production systems (MacRae and Mehuys, 1985). Legumes provide better soil fertility and enhancement of soil organic matter than cereal crops (Akobundu 1984; Agishi, 1985; Tarawali *et al.*, 1989).

2.3.4 Crop rotation

Crop rotation helps improve soil fertility and soil structure while controlling pests. Kamau *et al* (1999) carried out a three-year experiment comparing weed fallow with cow-pea rotation on the yield of maize both under conservation and traditional tillage systems. They found that cowpeas planted in the short rains season had a positive effect on the maize grain yield in the subsequent long rains season with both tillage systems. They attributed the effect to the contribution of nitrogen fixed by the cowpeas through nodulation and to the stalk and root residues left in the field after harvest. Similar studies have earlier demonstrated benefits of rotation with legumes (Gill *et al.*, 1992; Larney and Lindwall, 1995; Lenssen *et al.*, 2007).

2.4   Global trends in conservation tillage

Although, no-tillage has been in practice by farmers throughout the history of agriculture, application of herbicides as a way of abandoning traditional tillage started in the 1940s on large farms (Derpsch, 1998). Since then the practice has been expanding in different parts of the world with varying rates and by the year 2000, the total land under no-tillage reached 60 million hectares (Benites and Ashburner, 2001). Out of the total area under no-till, USA shares 21 million, Brazil 13.5 million, Argentina 9.3 million, Australia 8.6 million and Canada 4.1 million hectares (Benites and Ashburner, 2001). The realization of the detrimental effects of soil inversion using traditional plowing has influenced the rapid adoption of conservation tillage practices in North and South American countries (Derpsch, 1998).

2.5   Conservation tillage in Africa

Earliest research on no tillage in Africa was carried out in the late sixties in Ghana (Kannegieter, 1967; Ofori and Nanday, 1969). Research work at the IITA (International Institute of Tropical Agriculture) in Nigeria started in 1970 (Lal, 1983). Extensive on-farm experiments have also been conducted in the Eastern and Southern African countries (Steiner, 1998; Rockström *et al.*, 2001). Despite the wealth of research information on no tillage and mulch farming in Africa, the technology has hardly spread among farmers (Derpsch, 1998). Thus, no-till farming has mainly been practiced on commercial farms in South Africa and Zimbabwe (Steiner, 1998).

2.5.1   Major constraints

The adoption of conservation tillage is constrained by a number of factors that are both environmental and socio-economic. Castrignano *et al* (2001) stated that no single solution is universally suitable. The African Savannah lies between 10^0 and 16^0 N lat. (Rockström, 1997). The area is characterized by low and erratic rainfall ($300 - 600$ mm-yr^{-1}) coupled with high evaporation rates (1800-2300 mm-yr^{-1}), which makes it difficult to produce sufficient biomass for soil cover. Moreover, lack of grazing during dry seasons and the system of communal grazing restricts the possibility of leaving crop residues on the field after harvest (Kossila, 1988). On the other hand, in most studies where conventional tillage systems were compared with no-till, sufficient soil cover was maintained in the no-till treatments which prevented evaporation because of the insulating effect of the soil cover (Russel, 1939; Duley and Russel, 1939; Fuentes, *et al.*, 2003). Zero tillage without mulch can produce significant losses (Laryea *et al.*, 1991; Smith *et al.*, 1992; Rao *et al.*, 1998; Scopel and Findeling 2001). Ajuwon (1983) observed that in relatively low rainfall areas in Nigeria, zero-tillage systems were found to produce lower yields of maize and cowpea on topsoil with penetration resistance of >0.50MPa or with minimal earthworm activity. Moreover, in most studies where conventional tillage systems were compared with no-till, sufficient soil cover was maintained in the no-till treatments which prevented evaporation because of the insulating effect of the soil cover (Fuentes, *et al.*, 2003; Russel, 1939; Duley and Russel, 1939). The semi-arid regions in Ethiopia are characterized by low organic matter content (Mulatu and Regassa, 1986), which makes them prone to compaction. Under such conditions zero tillage without soil cover may not be relevant. Hence, in the African Savannah context CT is less about minimum tillage with mulch and more about wise tillage for water harvesting. The soils have been subject to degradation through traditional tillage with minimum bio-mass recycling for decades. Under such conditions, CT aims at reducing the quantity of tillage and at improving the quality of tillage. This means that no-till is not necessarily the objective. Instead we are focusing on tillage that maximizes infiltration with minimum soil evaporation.

One of the major challenges of conservation tillage is high weed infestations associated with reduced tillage. Rising fuel costs may make chemical weed control more attractive than tillage with tractors for commercial farmers. However, smallholder farmers use either manual labor or oxen for tillage, which often are much cheaper than the use of herbicides. Rutenbrg (1980) estimated that the energy required per ton of produce or per kilo calorie is less with ox ploughs than with tractors, for the simple reason that yield levels are the same for both systems, while the tractor system requires more horsepower hours. Moreover, many smallholder farmers are not skilled enough to properly handle and apply herbicides. The high costs of conservation tillage implements have also affected wider adoption of the technology (Steiner, 1998).

2.5.2   Adaptation of conservation tillage for smallholder farmers in Africa

In experiments to adapt CT systems to biophysical conditions it became clear that CT cannot be universally applied as a strictly defined practice, especially from the perspective of the agro-ecosystem of the African savannah. As a result, several types of tillage methods have been developed under the umbrella of conservation tillage (Rockström and Jonsson, 1999; Castrignano *et al.,* 2001; Diaz- Zorita *et al.,* 2002; Hoogmoed *et al.,* 2004).

For smallholder mixed farming system, a form of conservation tillage in which ripping along planting lines (animal traction) or planting in dug holes (hoe culture) and leaving the land in between undisturbed has been introduced among many farmers (Steiner, 1998; Biamah and Rockström, 2000; Kaoma-Sprenkels *et al.,* 2000; Kaumbutho, 2000; Nyagumbo, 2000; GART yearbook, 2001). Subsoiling with oxen drawn subsoilers has been an important component of the conservation tillage systems introduced in much of Africa because repeated tillage at shallow depth have resulted in the formation of plow pans beneath the plowing depth that restricted root growth and infiltration.

In Tanzania, trials carried out over a period of 3 years indicated that maize yields could be increased from 1.3 t ha^{-1} to 3.8 – 4.0 t ha^{-1} (Rockström *et al.,* 2001). In Kenya, slight increment in yield of maize was observed fom conservation tillage although the results were not statistically significant. In Zambia, deep ripping gave higher grain yields of maize although the differences were not statistically significant. Shallow ripping gave lower yields (Muliokela *et al.,* 2001). Boa-Ampongsem *et al* (2001) reported nearly double grain yield from conservation tillage systems using Roundup (Glyphosate) for weed control.

Several types of implements have been developed for conservation tillage. These include Magoye ripper and PALABANA subsoiler, weeders and direct planters introduced from Brazil and locally made in countries like Zambia and Zimbabwe (Kaumbutho and Simalenga, 1999; Hoogmoed *et al.,* 2003).

2.6   Conservation tillage in Ethiopia.
2.6.1   Reducing tillage frequency
For farmers who have traditionally used the mouldboard plow in conventional tillage, shifting to conservation tillage has usually involved replacement of the plow by rippers and subsoilers. However, since the mouldboard plow is not used by smallholder farmers in Ethiopia, the idea of reducing or minimizing soil

inversion has seldom been emphasized. The traditional tillage implement, the *Maresha* plow, does not completely invert the soil. Hence, the main concern has been the high number of tillage operations carried out with the *Maresha* plow. As a result, conservation tillage or minimum tillage in Ethiopia was regarded as reducing the number of tillage operations while using the same traditional tillage implement.

Many investigators (Taa *et al.*, 1992; Georgis and Sinebo, 1993; Tarekegne *et al.*, 1996; Tadele *et al.*, 1999) reported lower grain yields from minimum tillage treatments. Georgis and Sinebo, (1993) reported that in a trial conducted at Bako during 1982-86, maize grain yields were consistently higher with traditional tillage compared to minimum tillage. A three years trial conducted to study the effect of frequency of tillage on *tef* production at three locations in central Ethiopia showed that five times plowing with *Maresha* (the highest frequency) gave significantly higher grain yield compared to lower tillage frequencies (Tadele *et al.*, 1999). Taa *et al* (1992) compared four passes of *Maresha* with two passes and reported that wheat grain yield was significantly lower with two passes of *Maresha* compared to four passes. In another study conducted at Debrezeit, grain yield of wheat, when continuously grown for three years, was significantly reduced by the application of minimum tillage treatments (Tadesse *et al.*, 1994).

The reviewed experiments show that mere reduction of tillage frequency while using the same implement, the *Maresha* plow, can compromise grain yields thus making the practice unacceptable. Farmers have probably experimented with different tillage frequencies over centuries and have chosen the type of tillage frequency that they are using at the moment. Reducing tillage frequency would have been farmers' preference considering the limitations they face in terms of time, labor and traction requirement. Therefore, research has to come up with a different tillage system and/or different implements that can reduce tillage frequency.

2.6.2   Conservation tillage using herbicides
One of the purposes of tillage is weed control. With reduced or no tillage, weeds become a serious challenge. Use of non selective herbicides to control weeds before planting is the principal component of zero tillage. Conservation tillage systems that replace mechanical weeding by herbicide applications were introduced to Ethiopia by Sasakawa Global 2000 (SG2000) project (Gebre *et al.*, 2001). Extensive demonstration of such conservation tillage systems were conducted on farmers' fields. Erkossa *et al* (2006) conducted experiments on reduced tillage using herbicides for *tef* production in the highland vertisols reporting grain yield advantages of 8% over traditional systems. However, Gebre *et al* (2001) indicated that the cost of herbicides would be a concern in the adoption of the practice among smallholder farmers. The problem of affordability of herbicides as a major setback to the introduction of no-till system in smallholder farming system has also been reported by others (Ofori, 1993; Muliokela *et al.*, 2001). Moreover, the negative environmental effects of

herbicides, lack of skill for proper handling and application and costs of equipment for herbicide application remain to be challenges to the replacement of mechanical weed control by herbicides.

2.6.3   The issue of moisture conservation in semi-arid areas

Rockström and Jansson (1999) have shown that in the semi arid areas, water management is the key to improving crop productivity. Georgis and Sinebo, (1993) have also recommended the use of tied ridges and mulches in order to improve the soil moisture availability in the semi arid regions of Ethiopia. Whereas the use of herbicides can control weeds that are problematic with the introduction of reduced tillage system, leaving the soil undisturbed in the absence of cover crops or crop residues can result in high surface runoff. Hence, ways of opening the soil to allow infiltration while minimizing the adverse effects of tillage is a preferred strategy in developing conservation tillage systems for smallholder farmers in the semi-arid regions.

2.6.4   The case of broadcast crops.

In many sub-Saharan African countries, maize, which can and often is planted in lines, is the staple food. Maize is thus well suited for ripping based tillage in which only planting lines are cultivated while leaving the area in between undisturbed (Steiner, 1998; Rockström *et al.,* 2001). However, there has been very little effort to develop conservation tillage system for broadcasted crops. In Ethiopia, the majority of smallholder farmers grow a small seeded cereal called *tef (Eragrostis tef* (Zucc) Trotter). *Tef* cannot be planted in rows unlike maize and wheat because the plant has a very small size that makes it unable to fully utilize spaces left between rows. Hence, there is a need to develop conservation tillage systems that are suitable for *tef* production.

2.6.5   Implements

In Ethiopia, animal traction for tillage is associated with an ard plow known as *Maresha*. The *Maresha* plow has been used by the highlanders for thousands of years (Goe, 1987). It is very simple, light in weight, cheap, and is locally made. However, the *Maresha* Plow forms V-shaped furrows and results in incomplete plowing (Figure 3.2), which requires repeated tillage leading to a number of problems (figure 1.5).

Research to improve the traditional implement in Ethiopia, the *Maresha* Plow, began as early as 1939 when the Italians introduced the animal drawn mould board plow (Goe, 1987). The Food and Agriculture Organization (FAO) conducted a series of on-farm trials on implements in the 1950s while the Alemaya and the Jimma agricultural colleges made efforts to improve the traditional tillage implement in the early 1960s. In 1968, the Chilalo Agricultural Development Unit (CADU) started research on several types of tillage implements while the Institute of Agricultural Research (IAR) began activities on improving the traditional implements in 1974. However, none of these efforts were successful in developing prototypes acceptable by Ethiopian farmers (Goe, 1987). The major reasons behind the reluctance of farmers to adopt the newly introduced implements were the fact that they were too heavy and expensive (Goe, 1987). In order to develop acceptable implements for small scale farmers in Ethiopia, efforts were made to study and incorporate the design features of the traditional implement into the improved designs (Temesgen, 2001). Details of the improved implements are given in Chapter 4.

2.7   Hypotheses

The study has been conducted based on the following hypotheses.

1. Farmers undertake repeated tillage, for good reasons in terms of maximizing yields, based on resource availability and their perception about the purposes of tillage.

2. The improved tillage implements would perform the required tasks of timely operation, avoiding cross plowing and disrupting plow pans for conservation tillage.

3. Water productivity of maize and *tef* can be improved by using adapted conservation tillage systems. These include:

   a. A strip tillage system for maize in which only planting lines are cultivated can improve grain yields and water productivity. Moreover, use of subsoiler would maximize infiltration by breaking the plow pan so that runoff is reduced.

   b. Improved tillage systems for *tef* in which plowing is carried out only once with the *Maresha* Modified Plow followed by subsoiling and the use of the Sweep can control weeds and increase infiltration by reducing surface runoff thereby achieving higher grain yields and higher productivity than the traditional tillage system. .

Chapter 3

TRADITIONAL TILLAGE SYSTEMS OF SMALLHOLDER FARMERS IN SEMI-ARID AREAS OF ETHIOPIA[1]

3.1 Overview

In Ethiopia, a number of studies were carried out on tillage frequency in the high rainfall areas (Taa *et al.,* 1992; Taddele, 1994; Tarekegne *et al.,* 1996; Tadele *et al.,* 1999). Georgis and Sinebo (1993) reviewed studies carried out on tillage frequency in semi-arid regions. Their report favored repeated tillage with litle elaboration. Other investigators too (Mulatu and Regassa, 1986; Pathak, 1987; Beyene *et al.,* 1990) conducted surveys on the farming practices and implements in the central rift valley in Ethiopia, which is identified as semi arid (Engida, 2000). However, they did not analyze tillage practices in sufficient depth to obtain a clear understanding of the reasons for repeated tillage.

To assess the potential of conservation tillage, it is important to understand the reasons why farmers currently use conventional plowing methods. It is also necessary to determine if plow pans are formed under the traditional cultivation system so that appropriate measures such as subsoiling can be carried out.

This chapter presents the results of a study undertaken on traditional tillage systems in two selected sites in the dry semi-arid regions of Ethiopia with the objective of identifying reasons for repeated tillage and studying the presence of plow pans. The study relates tillage frequency with the type of implement used, type of crop grown, rainfall pattern and issues pertaining to the individual farmer such as skill of farming, resource availability, and perceived purposes of tillage. Moreover, the location, thickness and strength of the plow pan are presented.

3.2 Methodology
3.2.1 The study area
The study has been undertaken at Melkawoba and Wulinchity areas (Figure 1.3), which are typical dry semi-arid regions located in the central rift valley of Ethiopia. The two areas were chosen for their representations of the climates in the dry semi-arid regions (Engida, 2000). Within the dry semi arid category, Wulinchity is relatively wetter and with heavier soils than Melkawoba.

---

[1] Based on: Temesgen, M., Rockström, J., Savenije, H. H. G., Hoogmoed, W. B., Alemu, D. Determinants of tillage frequency among smallholder farmers in two semi-arid areas in Ethiopia. Accepted in: *Physics and Chemistry of the Earth.*

Melkawoba is located 08^0 23' North Latitude and 039^0 22' East Longitude with an altitude of 1450 m above sea level. The mean rainfall is 600 mm-yr^{-1} (Figure 3.1) with a potential evapotranspiration of 2300 mm-yr^{-1}. The rain is distributed over a period of 7 months (March-September) with two distinctive seasons (short rains in March and April are followed by the main rain season of June-September). The soil types are mainly sandy loam (*Calcaric Cambisols*) and very susceptible to compaction similar to the so called sealing, crusting and hard-setting (SCH) soils that are common in sub-Saharan Africa (Hoogmoed, 1999). Complete crop failure due to dry spells is not uncommon in the area. The major crops are *tef* (*Eragrostis tef* (Zucc.)) and maize (*Zea mays XX*).

Wulinchity is located 08^040' North Latitude and 039^026' East Longitude with an altitude of 1447 m above sea level. The soils are predominantly clay loam (*Eutric Cambisols*). The mean rainfall is 700 mm-yr^{-1} (Figure 3.1) while the mean potential evapotranspiration is 2200 mm-yr^{-1}. The rainfall distribution is similar to that of Melkawoba but usually sufficient rainfall is received during March and April to enable farmers to start tillage earlier. The types of crops grown at Wulinchity are similar to those of Melkawoba. Therefore, the study concentrated on *tef* and maize at both locations.

3.2.2   The traditional tillage implement

The traditional tillage implement, the *Maresha* plow, has been tested at both locations in order to observe the types of furrows it makes and the possible effects these may have on tillage frequency. About 15 parallel passes were made with the plow in exactly the same way as farmers plow their fields. The loose soil was then carefully removed by hand and the profile of the soil was examined to identify unplowed strips of land. The maximum depth of operation, width of the unplowed strip and that of the furrow were measured. A second tillage was carried out with the same plow laying furrows perpendicular to the previous ones and observations were made again. A third tillage was also carried out in the same direction as the first and the resulting soil profile was examined.

3.2.3   Studies on tillage systems

In order to balance between data reliability versus cost and time of research, 50 farmers were randomly selected from each area, out of 1450 and 1766 households at Melkawoba and Wulinchity, respectively. Randomization was done using the sampling frame of farm households' registration in each site and a selection interval to get 50 farmers was calculated and used in the process. A semi-structured questionnaire, which was pre-tested, was used to collect the primary data on farm household demographics, resource ownership, and perception about the purposes of tillage. Well trained enumerators who spoke the local language administered the questionnaire. The data were collected from June to September 2004. In addition to the questionnaires, group discussions were held with 10 farmers who were selected based on their farming experiences from each area.

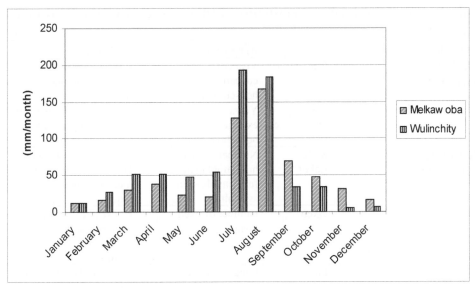

Figure 3.1. Average monthly rainfall at the study areas (1995-2004).

3.2.4   Analytical framework

The analysis was made using descriptive statistics and a regression model. Among farmers within a specified area, the number of tillage operations in any production system is hypothesized to be influenced by the skill of the farmer (years of farming experience and educational level), by the availability of production resources (labor, oxen and land), and by the farmer's perception about the purpose of tillage.

Different types of models (LOGIT, PROBIT and TOBIT) are used in situations where we have limited dependent variables (Sall *et al.,* 2000, Greene, 2003). However, TOBIT is more appropriate for limited dependent variables with values ranging from 0 to 1 while the other two are appropriate for limited dependent variables with choice of decision having values of either 0 or 1. In this study, the TOBIT model is used to identify the determinants of intensity of tillage in the production of maize and *tef* in the study area. The model is represented using an index function approach as follows:

$$Y_i^* = \max\left(X_i'\beta + \varepsilon_i, 0\right) \tag{3.1}$$

Where Y_i^* is an underlying latent variable that indexes the intensity of tillage; X_i is the vector of independent variables, containing: skill of farming, resource ownership, farmers' perception on the purpose and factors influencing intensity of tillage; β is a vector of parameters to be estimated; and ε_i is an error term. Table 3.1 shows description of factors that are hypothesized to influence tillage frequency. Y_i, the limited dependent variable, is the intensity of tillage calculated

Table 3.1. Hypothesized determinants of tillage intensity in *tef* and maize production

Variable		Measure	Rationale
Skill of farming	Education	Years of formal education of the household head	Educated farmers are able to judge optimum tillage intensity.
	Farming experience	Years of farming experience of the household head	Experienced farmers are able to judge optimum tillage intensity
Resource availability	Male Labor [2]	Number of male household members fully involved in agriculture	Households with enough labor can increase the intensity tillage as required
	Oxen	Number of oxen owned	Households with enough oxen can increase the intensity tillage as required
	Farm size	Land owned in ha	Households with larger farm size tend to specialize in crop production that allow them to allocate resources to increase tillage intensity as required
Farmers' perception about the purpose of tillage	Weed control	Weed control 1 = yes 0 = no	Farmers' perception about the role of tillage to control weed can influence tillage intensity positively
	Moisture conservation	Moisture conservation 1 = yes 0 = no	Farmers' perception about the role of tillage to conserve moisture can influence tillage intensity positively
	Manure incorporation	Manure incorporation 1 = yes 0 = no	Farmers' perception about the role of tillage to incorporate manure into the soil can influence tillage intensity positively
	Soil warming	Soil warming 1 = yes 0 = no	Farmers' perception about the role of tillage in warming up the soil can influence tillage frequency positively
Location		1 = Wulinchity, 0 = Melkawoba	Differences in locations such as soil types and rainfall can influence tillage intensity with heavier soils and more evenly distributed rainfall influencing tillage frequency positively

---

[2] Tillage in the study area is exclusively carried out by men.

as the ratio of each tillage frequency to the maximum tillage frequency recorded for each crop and each location. Hence, it ranges from zero to 1.

3.2.5 Studies on the plow pan

The penetration resistance of the soil was measured to a depth of 40 cm using manually operated cone penetrometer. Studies were undertaken on farmers' fields both at Melkawoba and Wulinchity. Tests were carried out during the dry season in order to minimize the effect of soil moisture on penetration resistance. Readings were taken at 18 randomly selected points in each of three fields (two maize and one *tef* fields) at both Melkawoba and Wulinchity.

3.3 Results and discussion

The socio-economic profile of sampled farmers is shown in Table 3.2. Compared to the national average land holding of 1 ha (Zekaria, 2002) households in the study area own more land (2.4 ha at Wulinchity and 1.4 ha at Melkawoba).

Farmers at Melkawoba are more educated but those at Wulinchity are older and with more years of experience in farming. Farmers at both sites allocated more land to *tef* than to maize. Farmers at Wulinchity own larger number of livestock, in general, and more oxen, in particular, which corresponded to larger land holding.

3.3.1 The traditional tillage implement

The traditional tillage implement, the *Maresha* plow, is shown in Figure 1.4. The *Maresha* plow forms a series of V-shaped furrows after the first tillage (Figure 3.2).

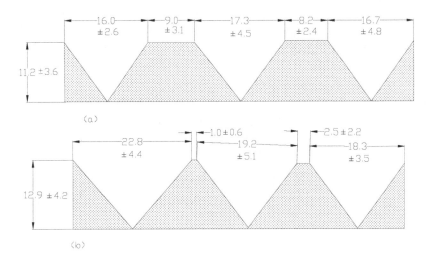

Figure 3.2. Profile of 3 furrows made after the first tillage by the *Maresha* plow at Wulinchity (a) and at Melkawoba (b). The profile was plotted after cleaning the loose soil. The shaded area represents the untilled part. The depth is the mean value of 15 readings while the width readings are mean values of 5 corresponding readings. All dimensions are in centimeters.

The furrows at Melkawoba were slightly wider and deeper than those at Wulinchity. The differences could be due to textural variations with soils at Melkawoba being lighter and more friable than those at Wulinchity.

Table 3.2. Socio-economic characteristics of sampled households (sample =100 farmers)

			Wulinchity		Melkawoba	
			Mean	STD[1]	Mean	STD
Socio-demographics	Age of household head (years)		44	14.2	40	15.1
	Education level of household head (year)		2.5	2.6	3.1	2.3
	Experience in farming (years)		28	14.2	23	12.8
Family labor	Family size	Total	5.4	2.4	5.3	2.7
		Male	2.8	1.2	2.8	1.6
		Female	2.6	1.7	2.5	1.7
	Number of household members with full time in agriculture	Male	1.5	0.68	1.6	0.56
		Female	0.5	0.74	1	0.47
	Number of household members with partial time in agriculture	Male	0.7	0.83	0.6	0.80
		Female	0.9	0.67	0.5	0.79
Land ownership and allocation	Land owned in ha		2.4	1.1	1.4	0.6
	Land allocated for maize (% of total)		21	15	30	18
	Land allocated for *tef* (% of total)		52	30	59	26
Livestock ownership	Number of tropical livestock units (TLU[2]) owned		3.8	2.4	2.3	1.9

Number of oxen owned	Percentage of farmers	
	Wulinchity	Melkawoba
0	8	26
1	14	42
2	49	28
>2	29	4

[1] Standard deviation.

[2] based on Kossila (1988). In the study areas animals owned by farmers and hence used for the calculation of TLU are chicken, goats, sheep, donkey, cows and oxen.

The V-shaped furrows made by the *Maresha* plow leave untilled land between adjacent furrows as shown in Figure 3.2. Farmers have to deal with the unplowed strips during subsequent tillage operations. The directions of any two consecutive tillage operations should be perpendicular to each other. Otherwise, the plow slips into previously made furrows thereby missing the untilled strip of land. When the plow is operated perpendicular to the previous tillage operation, it has to pass across the tilled furrows in order to reach the untilled ones. In so doing, an extra time of approximately 50% is spent on the already plowed land. Even after the second tillage, we observed spots of undisturbed land that required a third tillage during which even more than 50% of the time is spent on passing through already plowed land. In addition, repeated action on the soil causes excessive localized pulverization leading to structural damage. The other important effect of cross plowing is high surface runoff especially when plowing along moderate to steep slopes (Souchere *et al.,* 1998).

3.3.2   The traditional Tillage systems

Allocation of land: Land allocated to maize and *tef*, estimated by the farmers as a mean value for 10 years, is shown in Table 3.2. Farmers at both locations allocated more land to *tef* than to maize. The sizes of land allocated to *tef* and maize can vary with the rainfall distribution of a given season. Group discussions with farmers revealed that if sufficient rainfall is received in March and April, more farmers tend to grow maize especially the relatively long maturing varieties that give higher yields. If the rains start in the second and third week of June, fewer farmers plant early maturing maize varieties thus reducing the total land allocated for maize. However, if the rains are delayed until the last week of June or the beginning of July, they plant *tef* in moderate to steep slopes while beans are planted in the level fields of lower elevations. According to interviewed farmers, beans substitute maize in the level fields of lower elevations because *tef* is a poor competitor with weeds, which are more prevalent in these fields.

Table 3.3. Reasons why farmers do not plow before rains

	% of farmers	
Reasons	**Wulinchity**	**Melkawoba**
Too much draft force required	76	75
Too many clods formed	86	88
High weed infestation	88	88
Severe compaction by rain if dry plowed	49	88
To let weeds emerge after rains	60	67

Timing of tillage: Most of the farmers (88% at Melkawoba and 98% at Wulinchity) do not start tillage before it rains because of several reasons (Table 3.3). Soils are hard to plow when they are dry while oxen are weak during the dry season. Dry plowing forms too many clods that are difficult to break later thus resulting in poor seedling emergence (poor seed-soil contact). Moreover, repeated dry plowing results in excessive pulverization leading to crust formation and compaction. It is also desirable to let weeds germinate for a better control.

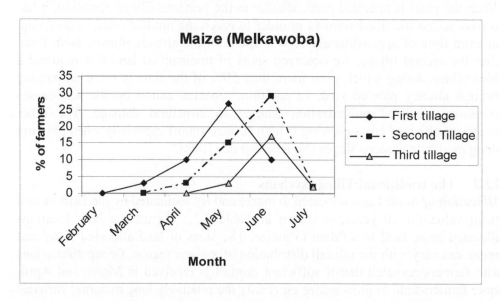

Figure 3.3 Timing of tillage for maize at Melkawoba (% of farmers).

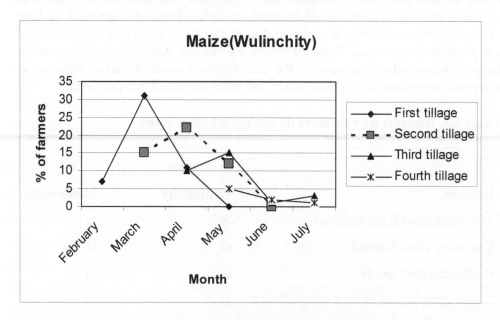

Figure 3.4 Timing of tillage for maize at Wulinchity (% of farmers). The graph shows tillage timing in seasons when rains start in February. In seasons with late onset of rainfall tillage frequency in the latter part of the season remains unchanged.

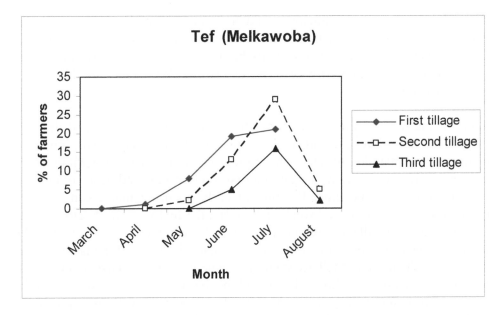

Figure 3.5 Timing of tillage for *tef* at Melkawoba (% of farmers)

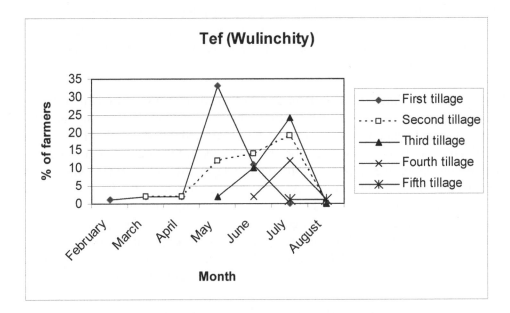

Figure 3.6 Timing of tillage for *tef* at Wulinchity (% of farmers)

Among the reasons given for not starting plowing before rains, major differences are observed between the two locations only in the case of compaction by rainfall. Compaction by rainfall is considered to be a lesser problem at Wulinchity probably because of the lower silt content of the soils. On the other hand, farmers realize that oxen sharing would be eased if they start plowing early because the period available for land preparation would be longer. Shortage of oxen was the reason why 2% and 12% of the farmers in Wulinchity and Melkawoba, respectively, plowed dry. The reason why more farmers plow dry at

Melkawoba than those at Wulinchity could be because of severe shortage of oxen at Melkawoba compared to that at Wulinchity. In other words, farmers who have to borrow oxen from others have to plow before the ox owners start plowing to avoid competition for oxen. Moreover, heavier soils at Wulinchity are difficult to penetrate when dry.

Tillage for maize can start as early as February (Figures 3.3and3.4), if there is sufficient rainfall to wet the soil. If the rainfall continues and the soil moisture can sustain crop, planting of late maturing maize varieties start in April or May. However, in many cases, rains start in the first week of June and hence tillage for maize starts at the same time. Planting of early maturing maize can then be carried out between the second week of June and the first week of August. There are differences in tillage commencement between the two locations. At Wulinchity, most farmers carry out primary tillage in March for maize and in May for *tef* while those at Melkawoba do the same in May for maize and in June-July for *tef* (Figures 3.3-3.6). This could be because there is more rain in February-May at Wulinchity than at Melkawoba as shown in the 10-year average rainfall distribution obtained from nearby stations (Figure 3.1).

Purpose of tillage: Purposes of tillage and their relative importance as reported by farmers are shown in Table 3.4. It can be observed in the table that the importance of a particular purpose of tillage varies with the stage of tillage (first tillage, second tillage, etc.).

During the first tillage, farmers intend to kill weeds that just started emerging as a result of preceding rain showers. Normally, farmers do not plow the land dry (Table 3.3) and hence weed seeds that are wetted by the first couple of rains start germinating. Tillage at this stage does not only control the germinated weeds but also brings more weed seeds from deeper layers to the surface which will germinate using subsequent rains. Asked what the purpose of the first tillage was, 87.5% of the farmers said it was to initiate weed germination while 74.5% said it was to kill the already emerged weeds (Average of maize and *tef* data in Table 3.4). The second and third tillage operations are thus aimed at killing weeds that emerged after the first and the second tillage, respectively. At planting, weed emergence is an inevitable consequence of tillage rather than a purpose.

Several researchers concluded that tillage causes loss of soil moisture. Baumhardt and Jones (2002) reported loss of soil moisture through evaporation due to tillage. However, they did not specify the time of plowing in relation to rainfall and dry spells. Osunbitan *et al* (2005) reported increased soil bulk density and hence reduction in hydraulic conductivity of plowed soils because of subsequent rainfall that occurred over a period of 8 weeks after plowing. Similar reports were also made by others (Mapa *et al.*, 1986; Fohrer *et al.*, 1999). However, the techniques used by farmers in the study area are different from the experimental set ups found in the reviewed literature. Farmers at Melkawoba and Wulinchity start plowing only after the rains start (Table 3.3). They carry out subsequent tillage operations following each set of wetting and drying cycle.

Plowing the soil two to three days after the rains stopped is convenient because of better friability. If plowing is delayed further, the soil will be too dry thus resulting in excessive pulverization that will lead to compaction by subsequent rain. According to farmers, tillage carried out at the right time reduces the bulk density of soils for increased infiltration from the following rains with less compaction and reduces evaporation thereby improving soil moisture.

Table 3.4. Farmers' perceived purpose of tillage at different stages (% of farmers)

Time of tillage	Purpose of tillage	Maize		*Tef*	
		Melkaw oba	Wulin chity	Melka woba	Wulinc hity
Primary tillage	Weed control	98	98	98	92
	Initiate weed germination	100	72	96	82
	Moisture conservation	100	84	98	82
	Soil warm-up	98	38	98	52
	Manure incorporation	60	18	28	58
Secondary tillage	Weed control	98	94	96	90
	Initiate weed germination	98	80	96	84
	Moisture conservation	98	86	98	80
	Soil warm-up	98	44	96	38
	Manure incorporation	60	56	50	60
Tillage at planting	Seed covering	100	100	72	62
	Weed control	100	68	98	66
	Initiate weed germination	60	36	98	66
	Moisture conservation	78	72	98	84
	Soil warm-up	62	48	28	24
	Manure incorporation	98	50	98	52

Guzha (2004) attributed the effect of tillage to increased soil roughness that increased surface area for water storage. Although, his results could be applicable in areas where evaporation is low, the farmers in the study area try to avoid increasing soil surface roughness in fear of higher evaporation losses.

The soils at both locations generally have low contents of organic matter (Mulatu and Regassa, 1986). Hence, manure application could improve soil fertility. However, only 62% apply manure on their field. The reason why the other farmers were not applying manure were, lack of manure (50%), unable to transport manure to the field (43%) and soil did not need manure application

(7%). The soil types that needed less manure were identified to be heavy textured. More than 50% of the interviewed farmers said that they plow their land in order to warm up the soil thereby improving seed germination and emergence. This needs further investigation.

Frequency of tillage: Factors affecting tillage frequency were found to be specific to the situation of a particular farmer, a community (location), soil type and environmental factors such as the distribution of rainfall.

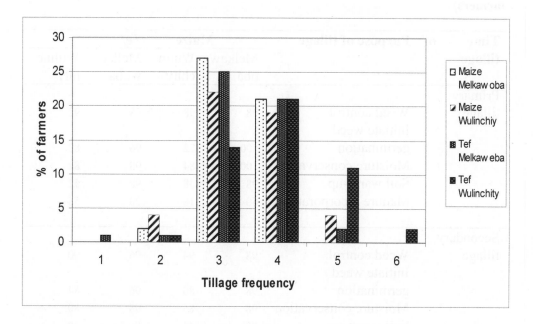

Figure 3.7. Tillage frequency for *tef* and maize.

Skill of farming, resources owned and perceived purpose of tillage affected tillage frequency in various levels (Tables 3.5 and 3.6). In these tables, the coefficient is a quantitative term describing the magnitude of the influence of the particular variable on tillage intensity while the t-ratio shows the level of significance of the effect of the variable. Education and experience had a significant effect in increasing tillage intensity. This could indicate that more intensive tillage is advantageous without considering the long-term effect it may have on soil quality which is difficult to be perceived by farmers irrespective of their level of education and experience. Tillage frequency for *tef* production increased with the number of male labor and farm size whereas in maize the effect was not significant. Higher tillage frequency was associated with larger farm size probably due to more specialization in farming instead of engaging in off-farm activities. Farmers who believe that tillage helps in warming up the soil tended to plow more frequently.

Farmers at Wulinchity plow more frequently (3.7 times on average) than those at Melkawoba (3.4 times on average) probably because of higher rainfall and heavier soils in the former. Higher rainfall is associated with more weed infestation. Moreover, the soils at Melkawoba tend to get excessively pulverized with repeated plowing.

Table 3.5 Determinants of the intensity of tillage in Maize production (TOBIT model)

Variables		Coefficient	t-ratio
Skill of	Education	0.0221	3.255***
farming	Experience	0.0046	3.222***
Resource	Male labor	0.0037	0.345
ownership	Oxen	0.0002	0.018
	Farm size	0.0191	1.012
Farmers'	Weed control	0.0992	2.047**
perception	Moisture		
about the	conservation	0.1381	3.158***
purpose of	Manure		
tillage	incorporation	0.0078	0.239
	Soil warm up	0.0623	1.721*
Location	(Wulinchity=1		
	Melkawoba=0)	0.0722	1.672*

Note: ***, ** and * represent levels of significance at 99%, 95%, and 90% probability.

Tillage frequency for *tef* is slightly higher (average 3.7) than that for maize (average 3.4, Figure 3.7). *Tef* fields are mostly plowed 3 to 5 times while 2% of the interviewed farmers said they plow 6 times. However, tillage frequency for *tef* in high rainfall areas is 5 to 9 (Tarekegne *et al.,* 1996). The difference could be because of the relatively low weed infestation in semi-arid areas as a result of low rainfall.

Moreover, farmers revealed during group discussions that *tef* is grown on moderate to steep slopes where weed infestation is lower than the level fields. Level fields are found in the lower elevations where relatively more moisture is available and where weed seeds are transported with run-on from the higher elevations. Since maize is more sensitive to moisture stress, it is normally grown in the lower fields where higher weed infestation forces farmers to plow more frequently resulting in the reported little difference in tillage frequency between *tef* and maize. Higher tillage frequency in *tef* could be more of the result of the need for finer seedbed. There is a much wider variation in tillage frequency in *tef* than in maize. This could be because of longer period of time available between onsets of rainfall and planting of *tef* in which case farmers decide whether to start plowing early depending on resource availability, soil type and level of weed infestation.

Table 3.6 Determinants of the intensity of tillage in *tef* production (TOBIT model)

Variables		Coefficient	t-ratio
Skill of farming	Education	0.0226	3.381***
	Experience	0.0023	1.675*
Resource ownership	Male labor	0.0249	2.488***
	Oxen	0.0156	1.157
	Farm size	0.0451	2.334**
Farmers' perception about the purpose of tillage	Weed control	0.0440	1.382
	Moisture conservation	0.1390	3.611***
	Manure incorporation	0.0391	1.366
	Soil warm up	0.0762	1.681*
Location	(Wulinchity=1 Melkawoba=0)	0.1949	4.798***

Note: ***, ** and * represent level of significance at 99%, 95%, and 90% probability.

According to interviewed farmers, rainfall that start early in the season and that are distributed over a longer period of time, with dry spells in between (Figure 3.8), lead to more frequent plowing. Farmers start plowing following the first two or three showers, which means early tillage commencement with early onset of rainfall. If this is followed by dry spells, the soil moisture gets depleted quickly and hence farmers decide to delay planting. Farmers carry out another tillage operation following a set of two or three showers in order to break surface crusts formed by raindrops impact and hence to improve infiltration. Moreover, farmers intend to minimize soil evaporation, locally known as *Nish Kebera* (hiding moisture), and to kill weeds that germinate following rainfall events. If such cycles of rainfall and dry spell continue without sufficiently wetting the soil for planting, farmers are forced to plow more frequently than when the rainfall comes late and continues without dry spells. Figure 3.8 shows how rainfall is distributed before planting period with wetting and drying cycles thus forcing farmers to plow frequently. In the years 2003 and 2005, the first set of showers started at the end of February. Farmers start plowing at this stage. In all the three years, there were couples of showers in the last week of March. Farmers plow the land two or three days after these showers. The dry spells that followed these showers deplete the soil moisture thereby forcing farmers to delay planting until another set of showers come. Such wetting and drying cycles are one of the reasons cited by farmers for increased tillage frequency before planting. It is also interesting to note that in 2005, the rainfall that occurred between the last week of April and that of May was high enough to enable farmers to plant maize in May. Such rare events encourage farmers to start plowing early in the season following the first set of rainfall although in many cases planting time is delayed until June (e.g. 2003 and 2004 shown in Figure 3.8).

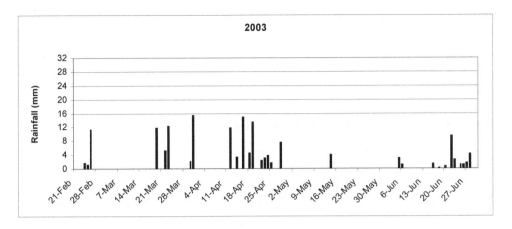

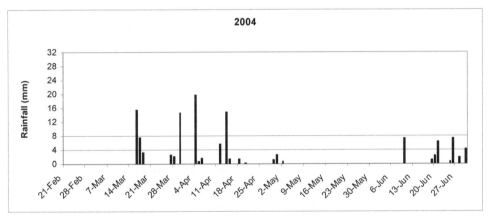

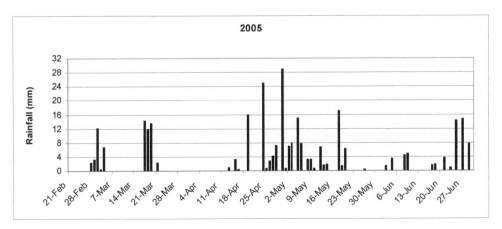

Figure 3.8. Daily rainfall and dry spell events before planting at Melkawoba (2003-2005)

The traditional tillage implement, the *Maresha* plow, forms V-shaped furrows (Figure 3.2) and leaves nearly half of the field untilled. This has forced farmers to plow more frequently since they have to carry out repeated cross-plowings to till the land completely and to control weeds.

3.3.3   Plow pans under the traditional cultivation system

The results of the field tests on penetration resistances of soils in cultivated fields are shown in Table 3.7, Figure 3.9 and Figure 3.10. It can be seen from the figures and tables that the maximum soil strength occurred over the depth range of 0.18 to 0.2m.

Table 3.7. Soil penetration resistance (kg-cm^{-2}, n=18)

| Depth (cm) | Melkaweba | | | Wulinchity | | |
	Field 1	Field 2	Field 3	Field 1	Field 2	Field 3
0	13	9	9	25	12	11
10	12	12	11	34	15	13
15	15	28	15	41	23	13
18	36	32	23	54	31	24
20	34	25	22	41	33	22
25	28	22	19	35	27	20
28	23	19	20	32	25	16
40	22	17	19	28	19	17

It is interesting to note that field 1 is where *tef* was planted the previous year. Farmers normally let animals trample over *tef* fields immediately after sowing in order to improve seed to moist soil contact for better emergence. As a result soils planted to *tef* are more compact than those planted to maize at both Wulinchity and Melkawoba. Field 2 and 3 are two different fields where maize was grown the previous year. It is also evident from the results that soils at Wulinchity especially those planted to *tef* are harder than those at Melkawoba. This could be due to textural differences, as determined earlier, soils at Wulinchity being heavier than those at Melkawoba.

Considering higher soil strength over the top 0.25 m layer it would be advisable to operate subsoilers up to and below 0.25 m in order to disrupt the plow pan created by the traditional *Maresha* cultivation system.

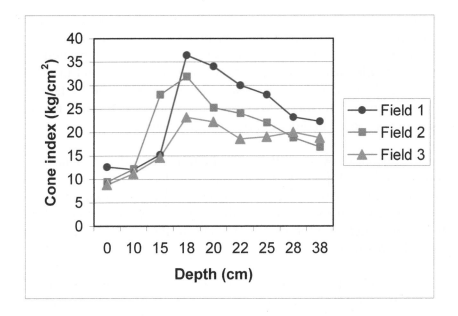

Figure 3.9 Penetration resistances at Melkawoba

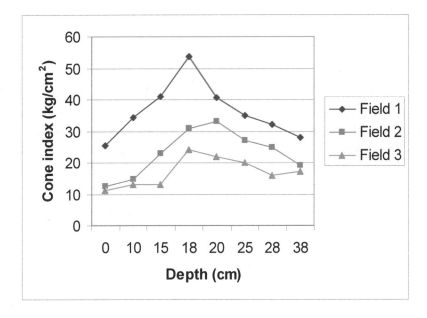

Figure 3.10 Penetration resistances at Wulinchity

3.4 Conclusions

The V-shaped furrows created by the *Maresha* plow leave unplowed strips of land between adjacent passes. Farmers are forced to undertake repeated and cross plowing in order to till the land left between furrows.

Commencement of tillage in the study area is dictated by the onset of rainfall. Farmers do not plow before the rain starts in order to minimize draft power requirements, to reduce soil pulverization thereby reducing compaction by subsequent rains, to minimize weed infestations, to reduce clods and to allow weed emergence for better control. Dry plowing is mostly practiced by farmers who do not own oxen and thus have to borrow from others before the normal tillage commencement period in order to avoid competition.

Tillage frequency increased with level of education and experience of farmers; size of farm owned; and perceived purposes of tillage such as moisture conservation and weed control. Tillage frequency also increased in heavy textured soils and in situations where rainfall occurred over longer periods of time. *Tef* fields are generally plowed 3 to 5 times while in most cases, maize fields are plowed 3 to 4 times.

The main purposes of tillage for growing maize and *tef* at Melkawoba and Wulinchity are soil moisture conservation and weed control. Farmers perceive soil warming as one of the purposes of tillage but this needs further investigation. Farmers in the study area plow the soil in order to improve soil water content through increased infiltration by breaking surface crust formed as a result of rainfall impact followed by drying and through reduced evaporation by closing evaporation paths, which is locally called *Nish kebera* (hiding moisture). Early onset of rainfall followed by repeated cycles of wetting and drying before planting time forces farmers to increase tillage frequency.

Plow pans are formed under the *Maresha* cultivation system. The plow pans are located at depths ranging from 0.18 m to 0.25 m. *Tef* fields at Wulinchity exhibited the highest soil strength while maize fields at Melkawoba had the least strength. It is recommended that subsoiling below the depth of 0.25 m be carried out in order to disrupt the plow pan and to allow better infiltration and root growth.

Chapter 4

MARESHA MODIFIED IMPLEMENTS FOR CONSERVATION TILLAGE[3]

4.1   Overview

Appropriate tillage implements are often the primary concern when shifting from conventional tillage to conservation tillage. In mechanized agriculture shifting to conservation tillage is usually accompanied by abandoning plows and harrows and introducing sprayers for the application of herbicides as it has been found that use of herbicides were more economical than mechanical tillage with tractors due to rising fuel costs (Derpsch, 1998). On the other hand, under smallholder farming systems use of herbicides for weed control has not been feasible (Ofori, 1993; Muliokela *et al.*, 2001). Under hoe culture, shifting to conservation tillage system requires the adoption of animal traction as the implements required for conservation tillage are mostly animal drawn. In Ethiopia, a large number of farmers use oxen for tillage (Pathak, 1987). However, the traditional oxen drawn implement, the *Maresha* plow, has been found to force farmers to undertake repeated and cross plowing, which causes land degradation (Chapter 3). Therefore, there is a need to come up with improved implements that can undertake conservation tillage systems without being too expensive, too heavy and too sophisticated for the resource poor smallholder farmers in semi-arid regions of Ethiopia.

Animal drawn implements developed to undertake conservation tillage under smallholder farming systems include the Palabama Subsoiler and the Magoye ripper (Jonsson *et al.*, 2000; Muliokela *et al.*, 2001). These implements were developed as attachments to the mould board plow frames that are too heavy and unaffordable by smallholder farmers in Ethiopia (Goe, 1987; Temesgen, 2000). In this chapter, different types of conservation tillage implements, which were developed as modifications or attachments to the traditional tillage implement, the *Maresha* plow, are presented. The results of testing their performance on farmers' fields are reported.

4.1.1   The *Maresha* plow - Opportunities and Drawbacks

Farmers in Ethiopia have used the *Maresha* plow (Figure 1.4) for thousands of years (Goe, 1987). It is very simple, light in weight, cheap, and locally made. However, as a conventional tillage implement, the *Maresha* plow has got several drawbacks which arise mainly from the fact that the plow forms V-shaped

---

[3] Based on: Temesgen, M., Hoogmoed, W., Rockström, J., Savenije, H. H. G. Conservation Tillage Implements for Smallholder farmers in Semi-arid Ethiopia. Accepted in: *Soil and Tillage Research Journal.*

furrows and results in incomplete plowing (Chapter 3). These drawbacks can have the following effects.

1. Because of incomplete plowing, farmers have to do repeated tillage in order to produce a fine seedbed especially for *tef*. As a result, the soil is excessively pulverized resulting in a poor structure (crust formation, compaction, etc.)

2. Because of the V-shaped furrow formed by the *Maresha* plow it is necessary that every two consecutive tillage operations are oriented perpendicular to each other. Thus, in inclined fields one of the two plowing operations fall along or nearly along the slope. The furrows encourage runoff when they are laid along the slope (Edwards *et al.,* 1993 and Basic *et al.,* 2001). This is a very serious problem in Ethiopia because the country is very hilly resulting in large amounts of soil and water loss.

3. Because of repeated tillage at shallow depth, plow pans are formed (Chapter 3), which hinder water infiltration (Whiteman, 1979) and root growth (Rowland, 1993; Willcocks, 1984).

4. The V-shaped furrow exposes a larger surface area of the soil to the atmosphere. Rough surfaces appearing during primary tillage operations enhance gas exchanges CO_2 (Reicosky, 2001) resulting in losses of organic carbon. Moreover, evaporation losses are higher due to larger surface area exposure.

Research to improve The *Maresha* plow began as early as 1939 when the Italians introduced the animal drawn mould board plow (Goe, 1987). FAO (Food and Agriculture Organization) conducted a series of on-farm trials on implements in the 1950s while the Alemaya and the Jimma Agricultural Colleges made efforts to improve the traditional tillage implement in the early 1960s. In 1968, the Chilalo Agricultural Development Unit started research on several types of tillage implements while the Institute of Agricultural Research began activities on improving the traditional implement in 1974. However, none of these efforts were successful in developing prototypes acceptable by Ethiopian farmers (Goe, 1987). The major reasons behind the reluctance of farmers to adopt the newly introduced implements were the fact that they were too heavy and expensive.

Opportunities: Although, the *Maresha* plow has these drawbacks, it also has a number of merits such as simplicity, lightness and low cost. Moreover, Rippers developed elsewhere such as those of the Magoye Ripper, cut the same type of V-shaped furrow as that of the *Maresha* plow. Hence, the *Maresha* plow can be used as a ripper without any modification. Other conservation tillage implements that were developed as modifications to the *Maresha* plow also are simple, light and cheap (Temesgen, 2000). This could be an opportunity to use these implements by resource-poor farmers. The following implements were developed as attachments to or modifications of the *Maresha* plow aimed at undertaking field operations related to conservation tillage.

4.1.2   The Subsoiler
Conventional tillage systems often cause the formation of plow pans or hard pans that restrict infiltration and root growth. One of the objectives of conservation

tillage practices is to break the hard pan. The implement meant for breaking the hard pan is called Subsoiler. Several types of subsoilers have been developed for animal traction including the Palabana Subsoiler. However, such implements were developed as modifications to the steel mould board plows that are made of heavy and expensive frames. It was, therefore, found necessary to develop a subsoiler as a modification of the *Maresha* plow. The Subsoiler (Figure 4.1) was made by replacing the tip of the *Maresha* plow by a narrow metallic part and the wooden boards by a pair of rods (Temesgen, 2000).

4.1.3 The Tie-ridger
In semi-arid regions, rainfall is erratic, which means a high intensity rainfall is followed by long dry spells. As a result, more water is lost through runoff and the soil dries out quickly during dry spells making less water available to the crop. Tied ridges are a series of basins formed in the field by tying furrows at certain intervals. They reduce runoff by creating obstacles to the movement of water thereby storing more water in the field and maintaining the soil moisture during dry spells. The equipment used to make such structures in the field is called a Tie-ridger (Figure 4.2). The Tie-ridger is made by attaching a blade to the tip of the Marehsa plow. The blade is designed to make wider furrows with reduced draft power requirement and lower lifting forces.

Figure 4.1. *Maresha* modified Subsoiler

Figure 4.2. The *Maresha* modified Tie-Ridger

4.1.4 The Sweep

The Sweep (Figure 4.3) is made by replacing the wooden boards of the *Maresha* plow by a pair of rods and a 55 cm wide blade. The blade cuts younger weeds that emerge after primary tillage. The main purpose of the Sweep is to improve timeliness of operations. It operates shallow thereby reducing draft power requirements and increasing speed of operation. It has got larger width of operation and hence more area is covered per unit of time. Moreover, the Sweep can be used to mix fertilizer with the soil during *tef* planting. Traditionally, farmers broadcast fertilizer on the surface and do not mix it with the soil. The reason is that if they mix it using the *Maresha* plow, the fertilizer will be buried too deep for the shallow-rooted *tef* crop. The drawbacks of leaving fertilizer on the surface include losses due to volatilizations and sheet erosion. Besides, the roots of the *tef* crop are forced to concentrate at the surface where the fertilizer is placed. The consequences are poor utilization of available moisture in the lower layer of the soil and lodging due to poor support. The former is particularly important when there is a dry spell during which the upper layer dries out with the moist layer progressively moving down the soil profile.

Figure 4.3. *Maresha* Modified Sweep

4.1.5   The *Maresha* Modified Plow

The *Maresha* Modified Plow (MMP) is the result of the incorporation of the bottom of the mould board plow with the frames of the *Maresha* plow (Figure 4.4). The incorporation of the bottom of the mould board plow into the *Maresha* plow is an opportunity to make the plow simple, light and cheap such that it is affordable by farmers in Ethiopia. The V-shaped furrow created by the *Maresha* plow has been identified to be the main cause of repeated plowing by farmers in Ethiopia (Chapter 3). It is, therefore, necessary that the implement be modified such that it forms U-shaped furrows that are more effective in weed control thus reducing the need for repeated and cross plowing. The reason for the V-shaped furrow created by the *Maresha* plow is the V-shaped geometry of its soil engaging part that penetrates deep at its tip while the wooden boards widen the furrow with the maximum width attained at the surface. MMP, on the other hand, operates with the full width of the share cutting at the bottom of the furrow. It is, therefore, hypothesized that MMP can form a U-shaped furrow which would then lead to reduced tillage frequency.

Figure 4.4. The *Maresha* modified plow.

4.1.6   The semi-automatic Row Planter.

Too much labor and time required for manual planting forces farmers to broadcast maize leading to uncontrolled and often too high plant density. Too high plant population leads to low water uptake per plant. Moreover, farmers do not apply fertilizer in broadcasted maize in addition to practicing poor tillage systems. The results are stunted growth and low yields (Figure 4.8).

Animal drawn Row Planters that were developed based on the designs of the tractor drawn Row Planters that involve ground wheels for metering seeds and fertilizer were not found to be effective under poorly prepared fields of the smallholder farmers. The rather light weight row planter equipped with small diameter wheels skidded and was unable to deliver seeds uniformly resulting in large unplanted gaps (Temesgen, 2000). A semi-automatic *Maresha* mounted Row Planter was developed in Ethiopia with a seed metering mechanism maneuvered by the operator instead of ground wheels (Temesgen, 2000).

The objective of this study was to evaluate the field performance of the newly developed implements with particular emphasis on their ability in reducing the time and energy required for the particular operation and in making more water available to the crop under rain fed agriculture in semi-arid regions.

Figure 4.5. *Maresha* mounted row seeder

Figure 4.6. The inverted BBM

4.2 Methodology

4.2.1 Testing sites

Field tests were carried-out in 2004 and 2005 at Melkawoba and Wulinchity. Full description of the sites is given in Chapter 3.

4.2.2 Implements tested

The implements tested were the Subsoiler (Figure 4.1), the Tie-ridger (Figure 4.2), the Sweep (Figure 4.3), the *Maresha* Modified Plow (Figure 4.4) and the Row Planter (Figure 4.5). The *Maresha* plow (Figure 1.4) and the inverted Broad bed maker (BBM) (Figure 4.6) were tested as checks. Tests were carried out on standard tests plots of 10 m x 40 m (Temesgen, 1995).

4.2.3 Measurement of draft power requirement.

Measurement of draft power requirement was carried out using a 5kN dynamometer for all the implements. The load cell was attached between the center of the yoke and the end of the plow beam. Readings were taken from a digital display attached to the load cell. Field performance tests were made on 40 m long plots for all the implements.

Testing of the Subsoiler: The Subsoiler was tested with due consideration of the way it will be used for conservation tillage. The type of conservation tillage designed for smallholder farmers in the semi-arid regions of Ethiopia involves ripping the field at a spacing of 0.75m, using the *Maresha* plow, followed by breaking of the plow pan along the ripped lines. Hence, the Subsoiler has been tested on furrows that were made by the *Maresha* plow. Comparisons were made with the use of the *Maresha* plow repeatedly along the same furrow to see if farmers can do without the Subsoiler. Thus, there were five sets of operations.

1. Single pass with the *Maresha* plow (*Maresha* alone)
2. Two subsequent passes with the *Maresha* plow (*Maresha* after *Maresha*)
3. Three subsequent passes by the *Maresha* plow (*Maresha* x 3)
4. A single pass by the Subsoiler over a furrow made by the *Maresha* plow (*Maresha* + Subsoiler)
5. Two subsequent passes with the Subsoiler over a furrow made by the *Maresha* plow. (*Maresha* + Subsoiler x 2)

Data were collected on depth of operation and draft power requirement. After each operation, the loose soil was carefully removed from the bottom and edges of the furrow at 10 randomly selected points along the furrow. A ruler was placed across the furrow and the maximum depth of operation was determined by measuring the vertical distance between the bottom of the furrow and the lower side of the ruler. The tests were replicated five times.

Additional tests were also made after the introduction of the closed furrow strip tillage system. In the latter, two or three adjacent passes were made using the *Maresha* plow. The Subsoiler was then used once at the middle of the plowed strip.

Testing of the Tie-ridger: The Tie-ridger has been tested in comparison with the *Maresha* plow and the inverted BBM. It was compared with the *Maresha* plow because we wanted to see whether farmers can undertake tie-ridging with the traditional implement. Moreover, the inverted BBM was tested because Bayu *et al* (1998) reported that it can be used for tie-ridging. Data were collected on draft power requirement and cross sectional area of the furrows produced. The latter was calculated after determining the maximum depth and width of the furrows. The cross sectional area of the furrow is related to the volume of water that can be retained in the basins created by the tie ridging operation. Ten readings were taken at randomly selected points at each furrow and the test was repeated five times.

Moreover, we measured the lifting force required when tying the furrows. A pocket balance was used to measure the lifting force with one end attached to the handle while the other end was lifted by the operator in the same way as the plow is lifted while tying furrows. A moving pen was attached to the indicator of the balance. The highest mark made by the pen was recorded as the lifting force. The mark was then cleaned every time a reading was taken. Ten readings were taken at randomly selected points at each furrow. The test was repeated five times.

Testing of the Sweep. Field performance evaluation of the Sweep has been made in comparison with the *Maresha* plow. Width of cut was measured using pocket meters by visually inspecting the soil cut by the Sweep without removing loose soils. Other data collected include draft power requirement, speed of operation, and total time required to complete the operations. The test was repeated five times and ten readings were taken during each test.

Testing of the Maresha* Modified Plow: The *Maresha* Modified Plow (MMP) has been tested in comparison with the *Maresha* plow. Data collected include draft power requirement, depth and width of cut, weeding efficiency and time requirement. Profiles of the soils were made at intervals of 2 cm both before and after plowing. Pegs were installed at the sides of the test plots on which a horizontal bar of 1.2 m long made of a rectangular hollow cross section of 50 mm x 20 mm x 3 mm was carefully placed. The bar has got 10 mm diameter holes drilled at a spacing of 2 cm through which 0.5 m long and 8 mm diameter rods passed. The rods moved freely through the section of the bar when placed on the ground both before and after plowing thus forming the profile of the soil. The heights of the rods protruding above the horizontal bar were measured to plot the profiles both before and after plowing. The depth and width of cut were determined using the readings thus taken. Weeding efficiency was determined by counting the number of weeds in a 1 m x 1 m steel frame that was randomly thrown at five spots in the test plots both before and after plowing. The difference between the number of weeds before plowing and that after plowing divided by the number of weeds before plowing and multiplied by 100 was taken as the weeding efficiency.

Testing of the semi-automatic Row Planter: Laboratory tests have been carried out to determine the number of seeds delivered per each stroke, which represents approximately 0.55 m span on the ground and to find out if the Row Planter missed delivery. The Row Planter was mounted on the *Maresha* plow and filled with maize seeds of *Katumani* variety. One person oscillated the wooden lever of the planter clockwise and counter- clockwise at a speed of approximately one stroke per second to simulate actual conditions in the field. Two people collected outputs from the delivery pipe on alternate strokes and counted the number of seeds thus collected while a fourth person recorded the readings. A total of 300 strokes were made with each stroke's delivery counted separately. Fertilizer was delivered in the same fashion after removing the seeds from the hopper. The total amount of fertilizer delivered for each set of 10 strokes was weighed using sensitive balances.

In the field, the Row Planter was compared with manual placement of seeds and fertilizer using the *Maresha* plow to open furrows. The Row Planter was adjusted to drop 3 seeds and 4 grams of Di-Amonium Phosphate (DAP) fertilizer per 0.5 m. The number of seeds placed in the hopper of the Row Planter was counted both before and after planting in order to determine the number of seeds placed in each plot. Plot sizes were 40 m x 10 m each so that 14 rows, with a spacing of 0.75 m, were contained in each plot. Manual placement was made behind the *Maresha* plow with 3 seeds and 4 grams of DAP fertilizer dropped at a spacing of 0.5 m. The number of labor and time required to cover each plot were recorded. Seedling emergence counts were made on the 7th, 14th and 21st day after planting. Grain yield and biomass data were also collected.

4.3   Results and discussion

4.3.1   Field performance of the of Subsoiler:

The results of the tests made on the Subsoiler are presented in Table 4.1. Statistical analysis shows that there are highly significant differences in performance in terms of working depth and required draft between the Subsoiler and the *Maresha* plow. However, differences between the respective third and second passes are not statistically significant.

The Subsoiler resulted in sufficient depth of penetration to disrupt the hard pan created under the *Maresha* plow. The maximum penetration resistance of undisturbed soils under the *Maresha* plow cultivation system was found to be located at a depth of 0.2 m (Chapter 3). This depth represents the hard pan which should be disrupted in order to allow infiltration and root growth. The *Maresha* plow was not able to break the hard pan even after it was used three times along the same furrow. On the other hand, the Subsoiler disrupted the hard pan in a single pass over a furrow made by the *Maresha* plow.

Two types of strip tillage systems, one with open furrow and the other with closed furrows have been tested (see Chapter 5) in order to minimize evaporation. The performance of the Subsoiler in the closed furrow strip tillage

Table 4.1. Field performances of the Subsoiler.

Treatment	Cumulative depth of cut (mm)	STD.	N	Change in depth[4] (mm)	Draft force[5] (kg)	STD	N
Maresha alone	154	1.2	50	0	107	12.4	96
Maresha x2	185	1.7	50	31	54	8.0	84
Maresha x3	207	3.8	50	22	49	3.9	90
Maresha + Subsoiler x1	239	0.7	50	85	52	4.4	79
Maresha + Subsoiler x2	268	4.5	50	29	51	5.0	81

system was significantly higher than in the single pass (open furrow) strip tillage system. Accordingly, the maximum depth of penetration of the Subsoiler, when used over a single pass of the *Maresha* plow (open furrow), was 0.23 m as opposed to 0.27 m when the Subsoiler was used over the middle of a strip made by two adjacent passes with the *Maresha* plow (closed furrow). The depth of penetration was even increased to 0.31m when the Subsoiler was used once over the middle of a strip made by three adjacent passes of the *Maresha* plow. The differences were statistically significant at P<0.01 (data not shown). Although, the depth of operation of the Subsoiler was increased to 0.27 m (see Table 4.1) by using the Subsoiler two times, which is comparable to the depth of operation when the Subsoiler was used once over two adjacent passes of the *Maresha* plow (0.27 m), the closed furrow strip tillage system was found to have other advantages over the open furrow (Chapter 5). Hence, it is recommended to use the Subsoiler once over two or three adjacent passes of the *Maresha* plow instead of using it twice over a single pass of the *Maresha* plow.

4.3.2 Field performance of the Tie-ridger.

The test results of the Tie-ridger are shown in Table 4.2. The larger cross sectional area of the tie-ridger compared to the *Maresha* plow and the inverted BBM (P<0.001) would enable retention of more water. As a result, rainfall partitioning can be positively altered by the Tie-ridger, making more water available for crop production in a semi-arid region where rainfall is erratic.

The draft power requirement of the Tie-ridger was also lower than the other two implements (P<0.001) which would enable the rather weak oxen in semi-arid regions to perform the job of tie-ridging with less energy. Moreover, the lower lifting force required by the Tie-ridger (P<0.001), when tying the furrows (Table 4.3), will considerably reduce the drudgery of the operation. The lifting force is so low with the Tie-ridger because the high inclination angle of the blade of the Tie-ridger reduces the vertical soil pressure that resists lifting.

---

[4] The change in depth refers to differences between two consecutive operations.

[5] The draft force was recorded for the last single operation.

Table 4.2. Draft force requirement and cross sectional area of furrows with different implements.

Type of implement	Draft force (kg)	N	STD	Cross sectional area of furrows (cm^2)	N	STD
The *Maresha* plow	103	118	7.1	305	50	30.8
The Tie-Ridger	79	78	6.0	416	50	38.5
The inverted BBM	96	85	6.9	363	50	46.9

Table 4.3. Lifting force required by different implements while tying furrows.

Type of implement	Lifting force (kg)	N	STD
The *Maresha* plow	42.3	50	6.6
The Tie-Ridger	24.6	50	3.5
The inverted BBM	43.9	50	4.8

The problem of moisture loss through soil evaporation observed with the timing of subsoiling (Chapter 5) is analogous to the problem reported on tie-ridging. This could partly explain the mixed results often reported on tie ridging. Several investigators (e.g. Hulugalle, 1990; Day *et al.,* 1992; Georgis and Sinebo, 1993; Bruneau and Twomlow, 1999) have reported positive results from tie ridging while others concluded that tie ridging was not effective under semi arid conditions (e.g. Vogel, 1993; Gicheru, 1994). The problems were in all cases related to the timing of the operation and linked to imbalances between allowing more infiltration and minimizing soil evaporation. Field trials should be initiated to evaluate the performance of tie ridging in relation to timing of the operation.

4.3.3 Field performance of the Sweep.

Table 4.4 shows the test results made on the Sweep to see its efficiency in terms of accomplishing light work, such as weed control and fertilizer incorporation during *tef* planting. The Sweep works faster than the traditional implement, because of wider operation and lower draft power requirement that enables the oxen to walk faster. It is expected that such a high work rate will enable dry land farmers to accomplish timely planting of *tef* that is critical to properly utilize the available growing period.

Table 4.4. Field performances of the Sweep.

Implement	Width (mm)	Speed of operation (m-s^{-1})	(ha-hr^{-1})	Draft force (kg)
Maresha	412	0.63	0.056	83
Sweep	562	0.71	0.083	52

Shallow operation by the Sweep can be attractive to farmers who face a dilemma whether to plow their maize fields during dry spells. Normally, after the initial tillage on maize fields, dry spells occur with a simultaneous emergence of weeds that exhaust soil moisture from the lower layers. Plowing the field with the *Maresha* Plow exposes the lower moist soil to evaporation. Leaving the lower soil layer undisturbed would maintain dry soil mulching at the surface, if farmers could control weeds otherwise. Such a dilemma could be dealt with by the use of the Sweep that just controls newly emerged weeds without exposing the lower moist soil.

4.3.4 Field performance of the *Maresha* Modified Plow.

The *Maresha* Modified Plow (MMP) took longer (27 hours-ha^{-1}) than the *Maresha* plow (23 hours-ha^{-1}) for the first tillage (Table 4.5). However, the *Maresha* plow left unplowed strips of land that had to be dealt with during the second and the third tillage operations (Figure 3.2).

On the other hand, MMP made U-shaped furrow thus leaving no unplowed land between consecutive passes (Figure 4.7). Therefore, repeated and cross plowing can be avoided by using MMP. Unplowed strips of land can be left intentionally at a chosen spacing in order to retard the movement of water depending on the slope of the field. The MMP was also superior to the *Maresha* Plow in weeding efficiency, depth of operation and width of cut.

Table 4.5. Field performance of the *Maresha* Modified Plow (MMP) at Melkawoba .

Type of implement	Draft force (kg)	Depth of cut (m)	Width of cut (m)	Weeding efficiency (%)	Time for first tillage (hr-ha^{-1})
The *Maresha* Modified Plow (MMP)	110	0.16	0.3	89.3	27
The *Maresha* plow (MP)	102	0.11	0.26	57.2	23
	NS	$P<0.05$	NS	$P<0.01$	$P<0.05$

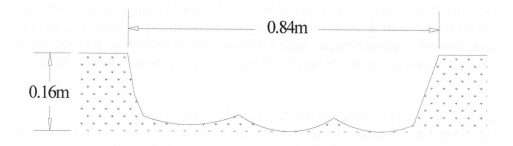

Figure 4.7. Profile of soil cut by three passes of the *Maresha* Modified Plow. Note complete disturbance of the U-shaped profile contrary to Figure 3.2.

4.3.5 Field performance of the Semi-automatic Row Planter.

Laboratory tests indicate that the Row Planter delivers 1-5 seeds per stroke with the maximum number of occurrence being 3 seeds (42%) and 2 seeds (31%). Four seeds were delivered 17% of the time while the shares of 5 seeds and 1 seed were 7% and 3%, respectively. No record of zero delivery was made during the 300 strokes.

Manual placement of seeds and fertilizer required three people (one operating the *Maresha* plow to open furrows, a second person to drop seeds and a third person to drop fertilizer) while the Row Planter was operated by one person only. The time required to complete a hectare of land using manual techniques (with three persons) was 26 hours-ha^{-1} while a single person required only 12 hours to do the same using the Row Planter with a covering device while 23 hours were required when the Row Planter was operated without the covering device. Thus, the maximum saving in man-hour (expressed as the product of the number of persons and time required to complete a given area) was 85%.

Table 4.6 Effect of use of Row Planter on seedling emergence (%) and grain yield (kg-ha^{-1}) of maize at Melkawoba.

Planting method	2004		2005		Mean	
	Planter	Manual	Planter	Manual	Planter	Manual
Seedling emergence (%)	78	42	71	46	75	44
Grain yield (kg-ha^{-1})	1300	1080	1670	1420	1485	1250

Table 4.6 shows the results of agronomic evaluation of the Row Planter on farmers' fields. The Row Planter resulted in significantly higher seedling emergence and higher grain yields of maize in both years (P<0.01). Higher seedling emergence was probably the result of better seed to moist soil contact because we observed that the Row Planter dropped seeds immediately after the furrows were opened by the *Maresha* plow as the planter was mounted directly on the latter. However, in the manual placement techniques, the person that drops seeds behind the *Maresha* plow is in most cases 5-10 m away form the plow and hence dry soil had the chance to flow back on to the bottom of the furrow, thereby covering the moist soil with a thin dry layer before the seeds are dropped. This may have resulted in poor seed to moist soil contact leading to reduced and delayed seedling emergence. Lower grain yields in the manual placement techniques are probably the result of delayed and reduced seedling emergence.

The combined effect of proper tillage system, fertilization and row planting produced up to 5 times the grain yield, compared to traditional practices of broadcasting seeds, improper tillage and no use of fertilizer (Figure 4.8). The maize crop on the right side was planted on the same date and with the same variety as the one on the left. The field is owned by the same farmer who has been cultivating his field in the same traditional method as in the one on the right side before the new systems were introduced. The crops in the experimental plots were easily distinguished from those in the rest of the fields by their vigorous growth and deep green color. In some years, farmers in the village did not harvest any maize at all, while the experimental fields still produced some. This scene drew the attention of many farmers who approached the author to learn more about the changes introduced. The author has been informed that some farmers have actually started practicing the new techniques during the main season of 2006.

4.4 Conclusions

The following conclusions can be drawn from these field experiments: The Subsoiler, when operated along furrows made by the *Maresha* plow, penetrated up to a depth of 0.24 m, which would enable disruption of the hard pan created under the conventional cultivation system of the *Maresha* plow. Moreover, a single pass of the Subsoiler over a strip that was cultivated with three adjacent passes of the *Maresha* Plow resulted in a 0.31m depth of penetration.

The Tie-ridger made furrows with larger cross sectional areas than those made by the *Maresha* plow and the inverted BBM while requiring lower draft forces. The lifting force required by the Tie-ridger when tying furrows was lower than that required by the *Maresha* plow and the inverted BBM.

The Sweep reduced the time and draft forces required during secondary tillage operations. Shallow operation of the Sweep can be used to control weeds during dry spells without exposing the lower moist soil.

The *Maresha* Modified Plow produced U-shaped furrows that made it possible to completely disturb the soil in a single operation thus avoiding the need for

repeated and cross plowing. Weeding efficiency and depth of operation were higher with the *Maresha* Modified Plow. The Semi-automatic Row Planter saved man-hour requirement of manual row planting, increased seedling emergence and increased grain yield of maize.

Economic analysis of some of these implements as related to conservation tillage systems are presented in subsequent chapters. The main reason for the economic benefits of these implements is the fact that they were developed as modifications and attachments to the *Maresha* Plow, which is locally made and inexpensive. Farmers felt greater level of technological ownership on these implements as they are acquainted with many of their components. The field tests have shown that these implements can be used, by resource poor smallholder farmers in Ethiopia, to undertake conservation tillage systems.

Figure 4.8. Picture comparing improved (left) and traditional (right) maize production systems. Changes made are tillage system, row planting and fertilization resulting in 5 times yield increment.

Chapter 5

WATER PRODUCTIVITY OF STRIP TILLAGE SYSTEMS FOR MAIZE PRODUCTION IN SEMI-ARID ETHIOPIA[6].

5.1.1 Overview

The traditional implement in Ethiopia, *Maresha* (Figure 1.4), and the tillage system that require repeated plowing have caused land degradation (Bezuayehu *et al.*, 2002), delayed planting and high drudgery to both draft animals and human beings (Pathak, 1987). Poor soil structure results in poor rainwater retention and infiltration (Rockström and Valentin, 1997; Hoogmoed, 1999) while delayed planting shortens the length of the growing period available for the crop (Rowland, 1993). Timeliness of operation is a serious problem for smallholder farmers in Ethiopia that cultivate 95% of the land under agriculture with more than 60% of them owning one or no ox (Pathak, 1987). Moreover, due to the geometry of the traditional tillage implement, farmers are forced to carry out cross plowing which orients the plow along the slope in one of two consecutive tillage operations thereby encouraging surface runoff (Chapter 3).

Incorporation of crop residues with tillage and repeated exposure of the soil to the atmosphere causes loss of soil organic carbon (SOC) through oxidation and mineralization. The rate of SOC loss upon conversion from natural to agricultural ecosystems is very high in the tropics (Lal, 2001). A number of studies have indicated decline of important soil qualities such as soil organic carbon and total Nitrogen (TN) following conversion of forest land into arable land (Lemenih *et al.*, 2005) although the rates could be variable (Lemenih and Itanna, 2004). SOC and TN improve soil quality and crop productivity (Bauer and Black, 1994).

Introduction of conservation tillage practices using appropriate equipment can help farmers improve soil quality for sustainable agriculture (Ahenkorah, *et al.*, 1995; Chen *et al.*, 1998; Steiner, 1998; Rockström and Jonsson, 1999; Biamah and Rockström, 2000; Freitas, 2000).

However, reduced or no tillage without soil cover results in reduced infiltration and lower grain yields (Georgis and Sinebo, 1993; Akinyemi *et al.*, 2003; Guzha, 2004). Such problems are inevitable in areas where lack of off-season rainfall and dry season feed shortage make it difficult to cover the soil either with crop residues or cover crops. This is typically the case in semi-arid Ethiopia, and this situation calls for an alternative approach. A strip tillage system may offer a solution.

---

[6] Based on: Temesgen, M., Rockström, J., Savenije, H. H. G., Hoogmoed, W. B. Water productivity of strip tillage systems for maize production in semi-arid Ethiopia. Submitted to *Agricultural Water Management*

Strip tillage systems where planting lines are cultivated while the inter-row space is left undisturbed have been found to have the benefits of both no tillage and conventional tillage (Mullins *et al.*, 1998; Lee *et al.*, 2003; Licht and Al-Kaisi, 2005). Moreover, strip tillage systems allow the farmer to plow only in one direction, along the contour, so as to prevent surface runoff. Tillage time is reduced thus enabling farmers to complete land preparation in time and reduce the oxen time required which would be particularly beneficial to resource poor farmers who own only one or no oxen at all.

This chapter reports on experiments evaluating a strip tillage system for maize production in two semi arid areas in Ethiopia. Traditional and improved tillage systems were evaluated for their impact on grain yield, soil water balance, and soil physical and chemical properties. Financial analysis was also made for the different tillage systems.

5.1.2 Materials and Methods
5.1.3 Experimental site
Experiments were carried out on selected farmers' fields at Melkawoba and Wulinchity. Detailed description of the two experimental sites has been given in Chapter 3.

5.1.4 Treatments
Three parallel treatments have been tested:
1) Conventional tillage (CONV) in which the land was plowed three to four times depending on the rainfall situation and according to farmers' practice.
2) Strip tillage system (ST) in which the planting lines were cultivated using the *Maresha* plow at 0.75 m spacing.
3) Strip tillage system with subsoiling (STS) in which the planting lines were cultivated using the *Maresha* plow followed by subsoiling with a *Maresha* modified Subsoiler (Figure 4.1) over the same furrows.

In 2003 and 2004, the design was a completely randomized block with 8 replications at each site whereas in 2005, ten replications were made. Each plot was 10 m by 10 m. A short cycle maize variety, *Katumani*, was planted in rows of 0.75 m spacing at a rate of 30 kg-ha^{-1}. In 2003 and 2004, the plots were split into subplots with and without fertilizer. Fertilizer was applied at a rate of 100 kg-ha^{-1} Di-Amonium Phosphate (23 kg N and 46 kg P_2O_5) at planting and 50 kg-ha^{-1} Urea 35 days after planting.

In the year 2005, all plots were fertilized. Moreover, due to early onset of rainfall in 2005 six blocks were planted with a medium maturing local maize variety called *Limat* on May 17, 2005. Four other blocks were planted with an early maturing maize variety called *Katumani*. Water balance studies were carried out in the two late planted blocks. Moreover, the furrows made along planting lines in the conservation tillage treatments were closed with a second pass adjacent to the previous as opposed to leaving them open. A separate experiment was also carried out comparing open and closed furrows both in STS and ST treatments to observe the effect of closing the furrows.

5.1.5 Water balance

Daily rainfall was measured at 9:00 o'clock using two rain gauges installed near the experimental plots. Pan evaporation, E_p, was measured daily at 9:00, 12:00 and at 15:00 hours using a Class A-pan installed near the experimental plots. Surface runoff was collected using a 0.5 m x 0.25 m x 10 m trough installed at the bottom of each of the 10 m x 10 m plot (Figure 5.1). The volume of water thus collected was manually scooped and measured using a 20-liter container and a graduated glass jar. Soil moisture was monitored using a Time-Domain Reflectometer (TDR) moisture measuring equipment from Eijkelkamp® and access tubes buried to depths of 1.8 m. Two tubes were installed on each plot in 0.04 m diameter holes drilled using hand augers. The holes were located 4.5 m from the North-West and South-East corners of each plot along the diagonal line that connects the two corners. With two replications, there were four tubes for each treatment. The mean values of data collected from the four tubes were used for the analysis.

Figure 5.1. Surface runoff collecting trough covered with hanging plastic sheets to prevent direct precipitation.

The leaf area index (I_{LA}) expressed as m^2-m^{-2}, was determined by measuring the maximum width and length of leaves on randomly selected 5 plants in each plot with a pocket meter at 30 and 60 days after planting. The leaf area (A) was calculated with the equation of Stewart and Dwyer, (1999). Thus,

$$A = \alpha W_M L \qquad (5.1)$$

where α is a coefficient with a value of 0.75 for the short stature maize variety used in our experiment, W_M is the maximum width of the leaf and L is the length of the leaf.

I_{LA} was calculated by adding the areas of all the leaves in each plant and dividing the sum by the area of land covered by each plant (Antunes *et al.*, 2001), which also means multiplying the total area of a single leaf by the population, P_0. Thus,

$$I_{LA} = P_0 \sum_{i=1}^{n} A_i \qquad (5.2)$$

where P_0 is plant population per m^2 and n is the number of leaves in each plant.

5.1.6 Modelling the water balance
Two types of models, a physically based and a conceptual model, were used to estimate the deep percolation and evaporation.

CoupModel
A physically based, one-dimensional ecosystems modelling package, the CoupModel (Jansson and Karlberg, 2004) was used to calculate evaporation of the maize crop for different tillage practices. CoupModel (and its precursor the SOIL-model) has been used in many climates and for several ecosystems (e.g. Blombäck *et al.*, 1995; Gärdenäs and Jansson, 1995; Alvenäs and Jansson, 1997; Rockström *et al.*, 1998; Gustafsson *et al.*, 2004; Karlberg *et al.*, 2005). In CoupModel, vertical water and heat transport in a layered soil profile is calculated with two coupled differential equations: the Richards equation for unsaturated flow and the Fourier law of diffusion (Jansson and Halldin, 1979). Net radiation is partitioned between the plant canopy and the soil according to Beer's law ((Impens and Lemeur, 1969). Precipitation infiltrates in the soil or forms a surface pool on the soil surface. In the CoupModel, no surface runoff is assumed, since measured surface runoff was deducted from the precipitation prior to the simulations. According to the Penman-Monteith equation (Penman, 1953; Monteith, 1965), total evaporation, E, is a function of net radiation and the vapour pressure deficit:

$$\rho_w \lambda E = \frac{\Delta R_n + \rho_a C_p \dfrac{(e_s - e_a)}{r_a}}{\Delta + \gamma\left(1 + \dfrac{r_s}{r_a}\right)} \qquad (5.3)$$

where R_n is net radiation available for evaporation (MJ-m^{-2}-day^{-1}), e_s is the vapor pressure at saturation (kPa), e_a is the actual vapor pressure (kPa), ρ_a is air density (kg m^{-3}), c_p is the specific heat of air at constant pressure (kJ-kg^{-1}K), λ is the latent heat of vaporization (kJ-kg^{-1}), ρ_w is density of water (kg m^{-3}), Δ is the slope of saturated vapor pressure versus temperature curve (kPa $^\circ$C^{-1}), γ is the psychrometer constant (kPa $°C^{-1}$), r_s is the surface resistance (s-m^{-1}) and r_a is the aerodynamic resistance (s-m^{-1}). The latter is calculated with a logarithmic function for wind speed:

$$r_a = \frac{\ln^2\left(\dfrac{z_{ref} - d}{z_o}\right)}{k^2 u} \qquad (5.4)$$

where the wind speed (m-s^{-1}), u, is given at the reference height, z_{ref} (m), k is von Karman's constant, d is the displacement height (m) and z_o (m) is the roughness length. Displacement height and roughness length estimated as fractions of plant height. Surface resistance, on the other hand, is calculated differently for the soil and the plant. The soil surface resistance, r_{ss}, is estimated as:

$$r_{ss} = r_{\psi 1}\left(\psi_s - r_{\psi 2}\right) \qquad (5.5)$$

where $r_{\psi 1}$ (s-m^{-1}) and $r_{\psi 2}$ (s-m^{-1}) are empirical coefficients and ψ_s (cm) is the water tension at the soil surface. Plant surface resistance for the canopy, r_{sc} (s-m^{-1}), is calculated as the inverse of the leaf area index of the plant, multiplied by the stomatal conductance, g_l. The latter is estimated with the Lohammar equation (Lohammar *et al.*, 1980; Lindroth, 1985):

$$g_l = \frac{R_{is}}{R_{is} + g_{ris}} \frac{g_{max}}{1 + \dfrac{(e_s - e_a)}{g_{vpd}}} \qquad (5.6)$$

where R_{is} is the incoming short-wave solar radiation, and g_{ris} (MJ m^{-2}), g_{max} (m s^{-1}), and g_{vpd} (Pa), are empirical parameters. Transpiration (mm-d^{-1}), E_{tp}, is reduced by a factor $f(\Psi(z))$ if the soil water tension in the root zone drops below a critical value, Ψ_c, according to:

$$f\left(\psi(z)\right) = \left(\frac{\psi_c}{\psi(z)}\right)^{p_1 E_{tp} + p_2} \qquad (5.7)$$

where Ψ is the soil water tension, z is soil depth, and p_1 (1 day^{-1}) and p_2 (kg m^{-2} day^{-1}) are empirical parameters. If water uptake is reduced from one specific soil layer, while other layers have a surplus of water, a compensatory uptake to meet part of the remaining demand for water will occur, as determined by the root flexibility degree. Finally, vertical drainage from the simulated soil profile of 2 m depth to lower soil layers is assumed to equal the unsaturated hydraulic conductivity at that soil depth.

Water flow was calculated using Darcy's law, modified by Richards (1931) to apply for unsaturated flow, and the law of mass conservation as:

$$q_w = -k_w\left(\frac{\partial \psi}{\partial z} - 1\right) \qquad (5.8)$$

where k_w (m-d^{-1}), is the unsaturated hydraulic conductivity, Ψ is the water tension and z (m) is depth.

Table 5.1. Parameters used in the coupmodel.

Property	Value	Unit	Source
Surface pool coverage, p_{max}	0.3	-	estimated from field observations
Reference height, z_{ref}	2	m	measured
Displacement, height factor, f_d	0.66	-	model default
Roughness length, height factor, f_{z0}	0.1	-	model default
Soil surface resistance coefficient 1, r_{Ψ_1}	10	s m^{-1}	Rockström et al., 1998
Soil surface resistance coefficient 2, r_{Ψ_2}	100	s m^{-1}	Rockström et al., 1998
Canopy conductance max, g_{max}	0.02	m s^{-1}	Nederhoff and de Graaf, 1993; Rockström et al., 1998
Canopy conductance RIS, g_{ris}	5	MJ m^{-2}	Heidmann et al., 2000
Canopy conductance VPD, g_{vpd}	1300	Pa	Heidmann et al., 2000
Critical threshold water uptake, Ψ_c	-400[*]	cm water	Values normally range from -100 to –3000 cm water.
Water uptake reduction coefficient 1, p_1	0.3	l day^{-1}	model default
Water uptake reduction coefficient, p_2	0.1	kg m^{-2} day^{-1}	model default
Root flexibility degree, f_{upt}	0.6	-	model default
Leaf area index, A_l	0-1.6** / 0-1.9/0-1.7	m^2 m^{-2}	measured
Plant height, H_p	0-1.8** / 0-1.9/0-1.9	m	measured
Root depth, z_r	0-1	m	measured
Extinction coefficient, k_{rn}	0.46 / 0.35	-	Kiniry et al., 2004
Plant albedo, a_{pl}	25	%	Oke, 1987; Gustafsson, 2002
Maximum surface coverage, p_{cmax}	0.75	m^2 m^{-2}	estimated from photographs
Surface coverage rate, p_{ck}	10	-	estimated from photographs

*A high value was chosen to represent a sandy loam soil.
**values for three treatments (CONV/STS/ST)

Daily climate data from a nearby climate station on air temperature, wind speed, duration of bright sunshine, relative humidity and precipitation for 2003-2005 have been used in the simulations. Precipitation has been modified for each treatment by subtracting runoff from the total amounts. The model has been parameterised based on field observations or literature values (Table 5.1). If neither of these were available, model default values were chosen. Soil texture data of a sandy loam soil measured in-situ was used to describe the soil in the simulations. Simulations were run from planting to harvest for each of the three seasons and for all treatments included in the study.

Conceptual model

A simple conceptual model (Figure 5.2) was also used to estimate the various components of the total evaporation (soil evaporation, transpiration and interception) and deep percolation. The model assumes that a certain proportion of the precipitation is intercepted by the canopy and soil surface, which is fed back to the atmosphere within the same day before it is partitioned between infiltration and surface runoff (Savenije, 2004).

A threshold D of 4 mm-d^{-1} was assumed for interception, resulting in the simple threshold function:

$$I = \min(P, D) \tag{5.9}$$

where I (mm-d^{-1}) is the evaporation from interception. The change in soil water storage was calculated using the water balance equation:

$$\frac{dS}{dt} = P - I - Q_s - T - E_s - R \tag{5.10}$$

where dS/dt (mm-d^{-1}) is the change in storage of water over the root depth (top 1m), P (mm-d^{-1}) is the precipitation, I is interception, Q_s (mm-d^{-1}) is surface runoff, T (mm-d^{-1}) is transpiration by the plant, E_s (mm-d^{-1}) is evaporation from the soil and R (mm-d^{-1}) is deep percolation below 1m.

When there is no limitation in soil moisture, plant transpiration is assumed to be related to the leaf area index, I_{LA}, (m^2-m^{-2}) the crop parameter, K_c, that also takes care of the pan coefficient, the pan evaporation, E_P (mm-d^{-1}), and I. Accordingly,

$$T_0 = I_{LA} \max(K_C E_P - I, 0) \tag{5.11}$$

where T_0 is the non-moisture-constrained transpiration. A value of 0.55 was assigned for K_C. However, when the soil water storage in the root zone, S, is below a certain value related to field capacity, S_{FC}, transpiration is reduced to a level determined by the curve that relates the ratio of actual transpiration to potential transpiration, T/T_0, with soil water storage, S (Figure 5.3).

The slope of the curve, K, is given by:

$$K = \frac{1}{(1-p)(S_{FC} - S_W)} \tag{5.12}$$

where $(1-p)$ is the fraction of soil water available to the crop $(S_{FC}\text{-}S_W)$ (mm-m^{-1}) in which transpiration is limited by moisture stress (Savenije, 1997). The ratio, T/T_O, is, therefore, given by:

$$T/T_0 = K(S - S_W) \tag{5.13}$$

Combining equations 5.11 to 5.13 yields,

$$T = I_{LA} \max\left(K_c E_p - I, 0\right) \max\left[\min\left(\frac{S - S_W}{(1-p)(S_{FC} - S_W)}, 1 \right), 0 \right] \tag{5.14}$$

Soil evaporation, E_S, is calculated using a similar concept as that of transpiration (Figure 5.4). The canopy cover will affect E_S and hence the area left uncovered expressed as $(1\text{-}I_{LA}C_C)$, is incorporated in the equation with the correction factor, C_C, to take account of leaf overlaps.

Thus, equation 5.14 can be modified into:

$$E_S = \max((1 - I_{LA}C_C)(K_S E_P - I), 0)\max\left[\min\left(\frac{S}{(1-r)S_{FC}}, 1 \right), 0 \right] \tag{5.15}$$

Water below the root depth is considered to be drainage, R, which is expressed as:

$$R = K_R \max(S - (1-p)S_{FC}, 0) \tag{5.16}$$

where K_R is a parameter that takes account of the share of deep percolation from storage in the root zone.

The change in storage is calculated using Equation 5.10. Subsequently, measured soil moisture content is compared with the simulated values to evaluate the accuracy of the simulations.

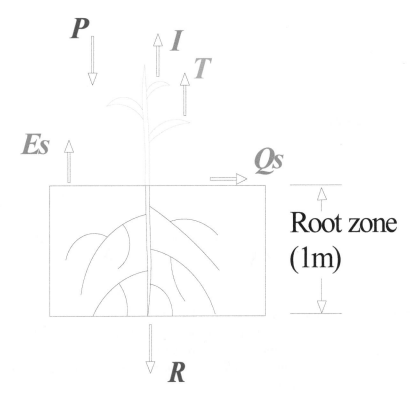

Figure 5.2. Conceptual model for rainfall partitioning in maize

5.1.7   Water productivity

Water productivity was calculated as the ratio of grain yield to total rainfall, total evaporation and transpiration using the formulae:

$$W_P = \frac{Y}{P} \qquad\qquad\qquad\qquad (5.17(a))$$

$$W_{PET} = \frac{Y}{\left(P - Q_S - R\right)} \qquad\qquad (5.17(b))$$

$$W_{PT} = \frac{Y}{T} \qquad\qquad\qquad\qquad (5.17(c))$$

where W_P, W_{PET} and W_{PT} are water productivity in kg-m^{-3} for total rainfall, total evaporation (the sum of soil evaporation, plant transpiration and interception) and plant transpiration, respectively, Y is grain yield in kg-ha^{-1}-season^{-1}, P is rainfall in mm-season^{-1}.

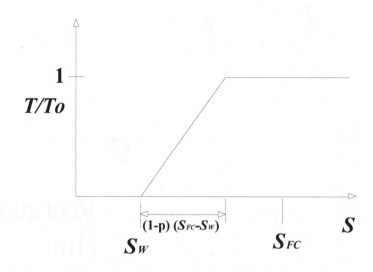

Figure 5.3. Ratio of actual and maximum plant transpiration (T/To) as affected by soil moisture. T: Actual plant transpiration, To: Maximum plant transpiration when there is no limitation due to moisture stress, S_W: Wilting point, S_{FC}: Soil moisture at field capacity and $(1-p)(S_{FC}-S_W)$: proportion of plant available water during which T is less than To.

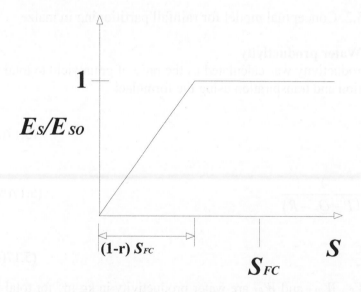

Figure 5.4. Ratio of actual and maximum soil evaporation (E_S/E_{SO}) as affected by soil moisture. E_S: Actual soil evaporation, E_{SO}: Maximum soil evaporation when there is no limitation due to moisture stress, S_{FC}: Soil moisture at field capacity and $(1-r) S_{FC}$: proportion of soil moisture during which E_S is less than E_{SO}.

Table 5.2. Parameters used in the conceptual model

Property	Value	Unit	Source
Crop coefficient, K_C	0.55		
Moisture content at field capacity, S_{FC}	17	%	Measured
Moisture content at wilting point, S_W	9	%	Measured
p	0.4		Assumed
r	0.3		Assumed
Interception threshold, I_D	4	mm-d^{-1}	Assumed
Leaf overlap factor, C_C	0.9		Field observation
Soil evaporation coefficient, K_S	0.5		Assumed
Drainage coefficient, R_C	0.03		Assumed

5.1.8 Grain yield

The crop was harvested leaving out one meter from each end and one row from each side of the plot. The total weight of above ground biomass was measured using a stationary balance of 20 kg capacity in the field. The cobs were carefully removed and shelled by hand and weighed. Moisture content of the grain was determined by drying in an oven at 70°C for 24 hours and grain yields were adjusted to a moisture content of 15.5%. Statistical analysis on the data was carried out using the SAS software (SAS Institute Inc., 1999).

5.1.9 Soil physical and chemical properties

Soil organic carbon (SOC), bulk density (BD), and total nitrogen (TN) and pH were measured before the experiment was started in May 2003. Samples were taken from the 0-0.15 m layer at 9 randomly selected points in the experimental field. Another 9 samples were collected up to a depth of 1.2 m from the same fields for textural analysis. At the end of the experiment, in November 2005, the same properties were measured from three randomly selected spots in each plot, after the crop was harvested.

The analyses were conducted according to the procedures outlined in Van Reeuwijk (1993). The particle size distribution (sand, 0.05 – 2.00 mm; silt, 0.002 – 0.05 mm; and clay, < 0.002 mm) was determined by the hydrometer method after organic matter removal. SOC was determined by the Walkley-Black method while TN was determined by the wet-oxidation procedure of the Kjeldahl method. The soils were classified according to the World Reference Base for Soil Resources (ISSS-ISRIC-FAO, 1998).

Table 5.3. Physical and chemical properties of soils at the initiation of the experiment at Melkawoba.

Soil texture	Sand	Silt	Clay
(Particle size distribution in %)	64	25	11
Bulk density (gm cm$^{-3)}$	1.36		
Organic carbon (%)	0.65		
Total Nitrogen (%)	0.08		

5.1.10 Financial analysis

An economic evaluation of the different tillage systems was made using the costs of tillage and weeding in each operation, and the gains from grain and stock calculated on basis of current market prices. Costs that were the same for all the tillage systems were not included.

The costs of implement use were calculated based on the current prices of the implements that are on sale while estimates were given to those not yet in the market (Table 5.11). Service lives were estimated for each tillage implement after consulting farmers as to how long the implements lasted in their respective areas while the time required for each operation were obtained from field measurements. The time required to complete the different operations in combination with data on service life was used to calculate the respective cost of implement use. In calculating the cost of implements, it is assumed that long-term credit will be made available to farmers by the government and hence no account of opportunity costs was made. Moreover, inflation has been ignored.

The costs of operation were calculated based on labor hiring rates of 8 Birr-day^{-1} (1 USD=8.7 Birr) and oxen hiring rate of 30 Birr-day^{-1}. Financial analyses were made by using the sum of costs of implement use, tillage operation and weeding as total expense and sales from maize grain and stocks as revenues. The price of maize grain was assumed to be 1.25 Birr-kg^{-1} while that of maize stock was assumed to be 0.25 Birr-kg^{-1}.

5.1.11 Participatory approach

Farmers were involved throughout the research period. At the beginning, workshops were held to introduce the concept of conservation tillage to farmers both at Melkawoba and Wulinchity. The treatments were proposed to farmers and amendments were made following their comments. The farmers were also encouraged to visit the experimental plots, to make observations and follow-up on the performance of the trial crops. In the following years, similar workshops were held to discuss the results of the previous year and to plan for the coming season.

5.2   Results and discussion
5.2.1   Water balance
Regression analysis of data on rainfall and surface runoff yielded the relationships shown in Equations 5.18-5.20. The figures in the bracket were used as thresholds to calculate the net rainfall.

$$Q_{S(CONV)} = 0.20(P-6.5)$$
$$R^2 = 0.7 \tag{5.18}$$

Where Q_S is surface runoff in mm-d^{-1} and P is rainfall in mm-d^{-1}.

$$Q_{S(STS)} = 0.09(P-5)$$
$$R^2 = 0.5 \tag{5.19}$$

$$Q_{S(ST)} = 0.13(P-6.5)$$
$$R^2 = 0.6 \tag{5.20}$$

Figure 5.5 shows the relationship between net rainfall and surface runoff.

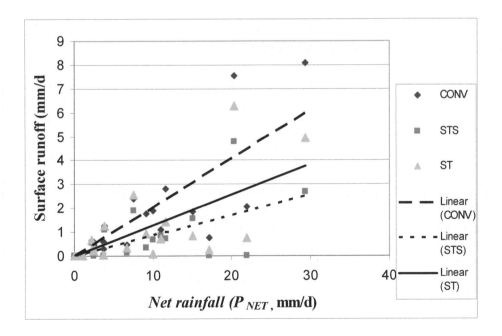

Figure 5.5. Surface runoff as affected by tillage systems (Melkawoba, 2005). P_{NET} is obtained after subtracting runoff threshold of each treatment from the total rainfall.

Surface runoff in CONV was the highest, probably because the loose soil did not resist the movement of water. We observed some rills in this treatment. On the other hand, the unplowed parts of the STS/ST treatments may have retarded the movement of water. Cross-plowing with the *Maresha* plow could also have increased surface runoff. Moreover, the subsoiled plots should have experienced more infiltration resulting in the lowest surface runoff. With reductions in surface runoff and hence possibly reduction in soil erosion, it is expected that the STS/ST treatments will have added benefits in the long term.

Model outputs

Tables 5.4 and 5.5 show the model outputs on water balance components. Table 5.4 shows outputs from the CoupModel. The rainfall has been adjusted after adding the runoff to the through fall values used in the model. The model initially used observed data on surface runoff but added extrapolated values for the dates between 27th of June (date when the CoupModel started the calculations) and 14th of July (date when the conceptual model started calculations) and when actual readings on surface runoff began. As it can be seen in Tables 5.4 and 5.5, the differences in rainfall given by the two models are large but the differences in surface runoff are negligible. This could be because the extrapolated runoff values were very small due to small rainfall readings recorded on several days between the 27th of June and 14th of July thus failing to generate significant amount of surface runoff. All other components of the water balance were outputs of the models.

Table 5.4. Effect of tillage treatments on water balance (mm-season$^{-1(7)}$)) at Melkawoba, 2005 and grain yield (Results from Coupmodel).

Treatment	P	Qs	I	T	Es	R	ΔS	T/P	Qs/P	$(I+Es)/P$
CONV	398	40	0	244	133	0	-21	0.61	0.10	0.67
STS	398	17	0	272	126	0	-17	0.68	0.04	0.63
ST	398	25	0	258	131	0	-17	0.65	0.06	0.66

Soil evaporation, transpiration and drainage: Model outputs, from both CoupModel and the conceptual model, show that STS has the highest transpiration to precipitation ratio followed by ST (See Figure 5.9 and Table 5.5). This is because of reduced surface runoff which makes more water available in the root zone. The ratio of the non-productive evaporation ($I+E_s$) to the precipitation is also the lowest in STS followed by ST.

---

[7] Season refers to period between sowing and harvesting.

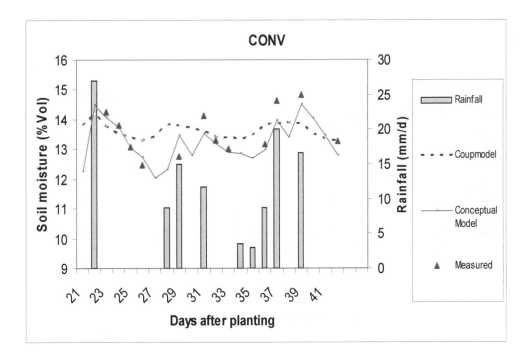

Figure 5.6 (a). Measured and simulated soil moisture over the root zone (0-1m) in CONV (Melkawoba, 2005)

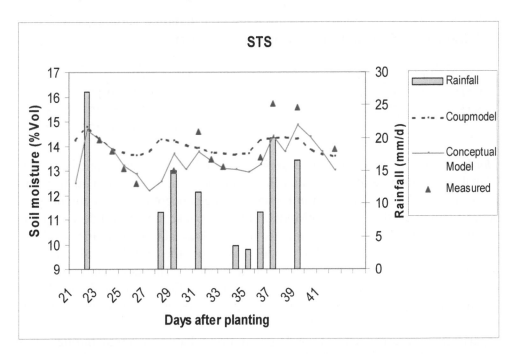

Figure 5.6(b). Measured and simulated soil moisture over the root zone (0-1m) in STS (Melkawoba, 2005)

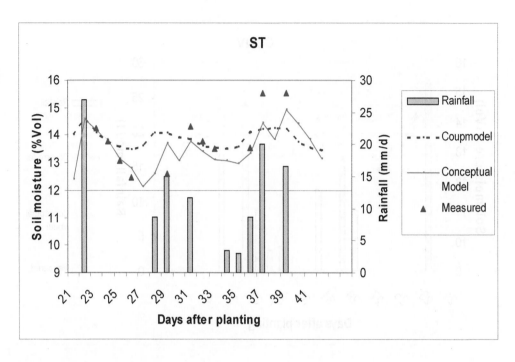

Figure 5.6(c). Measured and simulated soil moisture over the root zone (0-1m) in ST. (Melkawoba, 2005)

Table 5.5. Effect of tillage systems on water balance (mm-season$^{-1(8)}$)) at Melkawoba, 2005. (Results from conceptual model).

Treatment	P	Qs	I	T	Es	R	ΔS	T/P	Qs/P	$(I+Es)/P$
CONV	355	40	100	158	39	56	-37	0.44	0.11	0.39
STS	355	17	100	196	25	54	-36	0.55	0.05	0.35
ST	355	25	100	178	31	57	-36	0.50	0.07	0.37

The hydrology of the unsaturated zone: Flow in the unsaturated zone is the sum of matrix, diffusion and preferential flows. In the CoupModel, only matrix flow is considered. As it can be seen in Figure 5.6, the conceptual model outputs gave better simulations than the CoupModel. This is because the general description of the unsaturated zone as a continuous matrix does not hold in the semi-arid tropics. As a result a complex model based on the full Richards equation performs worse than a simple conceptual model. The dominant infiltration mechanism in the unsaturated zone has been identified to be preferential flow (Gjettermann *et al.,* 1997; McGlynn *et al.,* 2002) especially in dry situations with erratic rainfall (Ritsema and Dekker, 2000). In the sandy loam soil where the experiments were carried out, preferential flows likely occurred through dead plant roots (Perillo *et al.,* 1999) that are less disturbed by the rather shallow tillage (Petersena *et al.,* 2001) with animal traction. In the simple conceptual model, the one layer approach averaged the preferential flow effect and thus the

---

[8] Season refers to period between seedling emergence and harvesting.

model output is better comparable with field measurements. Figure 5.7 is an illustration of a preferential flow system.

Table 5.6. Effect of tillage system on grain yield of maize (kg-ha^{-1}).

Treatments	Melkawoba					Wulinchity		
	2003	2004	2005[9]	2005[10]	Mean	2003	2004	Mean
CONV	1390	1070	2100	1720	1570	1170	1610	1390
STS	1430	920	1650	2130	1530	1200	1480	1340
ST	1520	1010	2000	1840	1590	1170	1600	1380
	NS	NS	NS	NS		NS	NS	
Rainfall (mm-yr^{-1})	674	497	688	688		786	580	

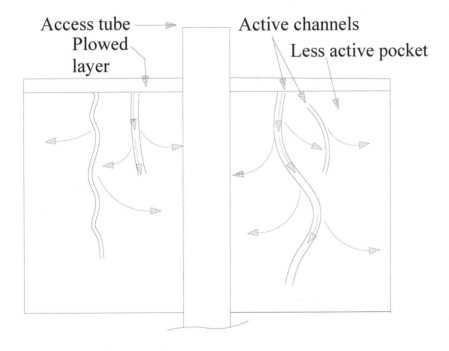

Figure 5.7 Preferential flow occurring primarily through active channels. Moisture redistributes in the less active pockets at a later stage.

Infiltration takes place rapidly along the active channels and then starts redistributing horizontally to the less active pockets. This phenomenon can also be observed in the TDR readings. The curves in Figure 5.8 represent different dates during which TDR readings were taken. On July 22nd (26 days after planting), the soil was relatively dry after 5 days of dry spells. Following a 24

---

[9] Medium maturing maize variety, Limat

[10] Early maturing maize variety, Katumani

mm rainfall event, the moisture in the upper layer (0-20 cm) increased from 12.6 to 15.5% while in the other layers it remained unchanged. The effect of redistribution of moisture from the active channels to the less active pockets took place 3-4 days later as seen on 27-July. This explains why the measured soil moisture was less than the predicted on July 25. On July 29, after two days of dry spell, we find uniform soil moisture obviously due to the fact that drying takes place uniformly. The drying process has been simulated better by the conceptual model (Figure 5.6).

5.2.2   Grain yield

Monthly rainfall distribution over the experimental seasons is shown in Figure 5.10. Generally 2004 was a relatively dry season whereas the distribution and the amount of rainfall in 2005 were better than those in the other two seasons. The grain yields increased with the amount and distribution of rainfall. The results are shown in Table 5.6. The differences in grain yield among the different tillage systems are not statistically significant.

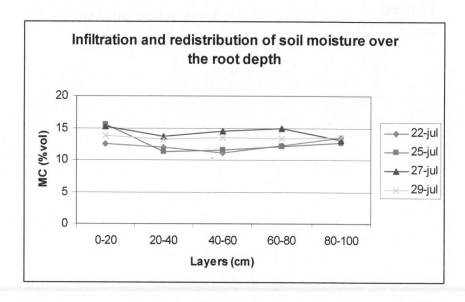

Figure 5.8 . Infiltration and redistribution of soil moisture computed over the root depth (0-1 m) as affected by rainfall and dry spells. 22-Jul: after 5 dry days; 25 July after 24 mm; 27-Jul a dry day followed by 12 mm rainfall; 29 Jul after 2 dry days.

It is interesting to note that in the year 2005, the results were in favor of conventional tillage for medium maturing maize variety while the conservation tillage treatments gave higher yields for the early maturing maize variety. For the medium maturing maize variety, participating farmers commented that the soil in STS/ST treatments lost more moisture because of higher soil evaporation than in CONV. This was because the time between the date of the last tillage in STS/ST and planting (DTP) was longer (48 days), in the case of the medium maturing variety, while DTP was 6 days in the case of the early maturing variety (Figure 5.11). Reasons for delayed planting are described in Chapter 3 Section 3.3.2.

The effect of higher DTP could be higher loss of soil moisture due to transpiration by weeds as the latter were not controlled, more soil evaporation and more compaction by rainfall during the extended periods that reduced infiltration of subsequent rainfall. Motavalli *et al* (2003) reported that re-compaction of loosened subsoil due to deep tillage may be relatively rapid, thereby limiting its effectiveness over time. Figure 5.11 also reveals that STS was more sensitive to DTP than ST. This could be because of higher infiltration followed by higher soil evaporation in STS due to the effect of subsoiling. Similar trends were also observed in the results of the previous years. In 2003, DTP was 26 and 23 days at Melkawoba and Wulinchity, respectively, while in 2004 there was a 59 and 56 days gap at Melkawoba and Wulinchity, respectively, which could have lowered grain yield from ST/STS relative to CONV in 2004 as compared to that of 2003.

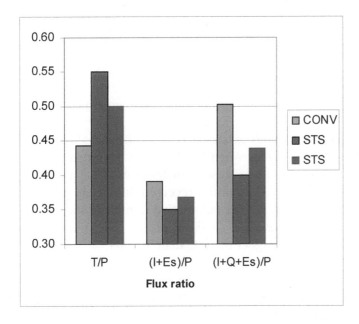

Figure 5.9 Ratio of productive (transpiration), non productive (interception, soil evaporation) and runoff to total precipitation as affected by tillage systems.

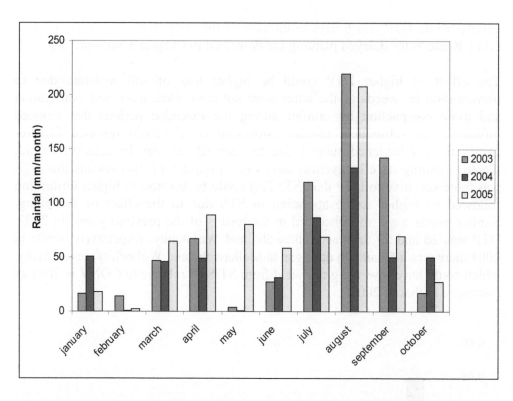

Figure 5.10. Monthly rainfall at Melkawoba during the experimental years.

When comparing the two situations with extended DTP of 48 and 59 days, we find that the effect of the gap was higher when DTP was 48 days than when it was 59 days (Figure 5.11). This could be because of differences in the amount of rainfall received during the two gaps. The rainfall received during the gaps was 105 mm in 2005 and 65 mm in 2004. Higher cumulative rainfall in the 48 days gap could have caused more compaction (Ndiaye *et al.*, 2005) and more evaporation. The grain yields are also higher in 2005 than in 2004.

The results indicate that in future studies, it may be necessary to cultivate the planting lines in STS/ST treatments at a shallow depth using the Sweep (Chapter 4) in situations where longer periods between tillage

Table 5.7. Effect of fertilization on grain yield of maize.

Treatment	Melkawoba		Wulinchity	
	2003	2004	2003	2004
Fertilized	1479	1146	1317	1668
Un fertilized	1408	860	1040	1461
	NS	P>90%	P>95%	P>95%
Fertilizer x Tillage	NS		NS	NS

commencement and planting cause soil compaction by rainfall. The Sweep could help in controlling weeds thereby reducing weed transpiration while at the same time breaking crust for increased infiltration and sealing vertical channels for reduced evaporation, without exposing the lower moist soil layers. Moreover, the option of late subsoiling such as one week before planting or after planting should be tested.

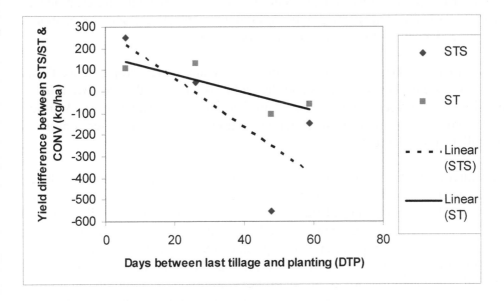

Figure 5.11 Performance of conservation tillage systems in relation to days between last tillage and planting (DTP)

Table 5.8. Effect of closing furrows in STS/ST on grain and biomass of medium maturing maize (Melkawoba, 2005)

Treatment	Grain yield (kg-ha^{-1})	Biomass (kg-ha^{-1})
Open STS	993 (c)	3576 (c)
Open ST	1250 (bc)	4347 (bc)
Closed STS	1332 (ab)	4875 (ab)
Closed ST	1587 (a)	5750 (a)

Means followed by the same letter are not statistically different

Closing planting furrows in STS/ST treatments showed significantly higher grain yield compared to leaving them open (Table 5.8). The reason could be higher loss of moisture in the open furrows over the planting zone. Open furrows could reduce surface runoff during heavy storms by acting as barriers to downward movement of water but during dry spells they can cause higher soil evaporation.

During 2003 and 2004, fertilizer was applied in split plots. Although the effect of fertilizer was significant in both years and locations except in 2003 at Melkawoba, low soil moisture in the study areas could be the reason for the rather little response to fertilization. There was no interaction between tillage system and fertilization again probably because the new tillage systems did not make appreciable changes in soil moisture during 2003 and 2004 as explained earlier.

5.2.3 Water productivity

In a broad sense, productivity of water refers to the benefits derived from a use of water. Water productivity is dependent on several factors including crop genetic material, water management practices, agronomic practices, and the economic and policy incentives to produce (Kijne *et al.*, 2003).

Water use and management in agriculture cross many scales: crops, fields, farms, delivery systems, basins, nations and the globe. Working with crops, we think of physiologic processes: photosynthesis, nutrient uptake and water stress. At a field scale, processes of interest are different: nutrient application, water-conserving soil-tillage practices, etc.

Considering the banner of "more crop per drop" or "producing more with less water" we find that different people define the word 'drop' in different ways such as more kilograms per unit of transpiration by breeders and more kilograms per unit of rain water by agronomists and agricultural engineers. In addition to selecting crops and crop varieties that produce more with less water, farmers in rain-fed arid areas are concerned with capturing and doing the most with limited rainfall. By capturing we refer to proper partitioning of rainfall into useful components such as infiltration as opposed to surface runoff. One may argue here that surface runoff can be used by downstream users. However, surface runoff occurs during the rainy season at a time when there is sufficient rainfall and hence downstream users benefit less from the surface runoff. Moreover, in addition to losses of soil and nutrients by the farmer, the negative effect of surface runoff extends to polluting rivers and silting dams. On the other hand, infiltrated water is further divided between transpiration and drainage. The drainage component contributes to the downstream flow with much slower speed than surface runoff thereby contributing to downstream flow at a time when there is water scarcity such as during the dry season.

Coming to the objectives of conservation tillage, which is minimizing surface runoff by maximizing infiltration, the best way of evaluating the contribution of newly introduced conservation tillage systems would be to look at what proportion of the rainwater has been used for crop production. In other words, in calculating water productivity from the point of view of conservation tillage the denominator should be the total amount of rainwater that was partitioned between surface runoff and infiltration. A tillage system that resulted in more infiltration will show higher water productivity since it has made it possible to use higher proportion of rainwater for crop production.

Accordingly, Table 5.9 reveals that in the year 2005, when we had better performance of the conservation tillage treatments particularly, STS, we find that water productivity for total rainfall (W_{PP}) showed the highest value for STS thus reflecting on the performance of the tillage system. As explained above, this may have stemmed from the fact that STS had the least surface runoff, which means more of the rain water was used for crop production. It is evident from Table 5.9 that water productivity for transpiration (W_{PT}) did not show appreciable differences among the treatments. Had we compared different crops or crop varieties we could have found significant differences in W_{PT} because for the same amount of transpired water, different crops or different crop varieties would give different grain yields. However, water productivity for total amount of evaporation (W_{PET}) appeared to be a better way of assessing water productivity than transpiration alone. This could be because of vapor shift (Rockström, 2003), which reduces soil evaporation as biomass production increases in water conserving treatments as a result of reduction in surface runoff. The increased W_{PET} as a result of the vapor shift leads to a more efficient use of the depleted (evaporated) water in STS than in CONV. This will have a positive contribution to the water productivity on both the catchments and basin scales.

Table 5.9 Water productivity (kg grain m^{-3})as affected by tillage systems in maize

	Melkawoba 2005		
Treatment	W_{PT}	W_{PET}	W_{PP}
CONV	1.09	0.58	0.48
STS	1.09	0.67	0.60
ST	1.04	0.60	0.52

5.2.4 Soil properties

Tillage treatments did not significantly alter soil physical and chemical properties after a period of three years (Table 5.10). According to some literature, the SOC and TN contents of soils take longer (>5 years) to respond to reduced tillage (West and Post, 2002; Heenan *et al.*, 2004) while others reported significant changes in shorter periods of two to three years (Su *et al.*, 2004; Ozpinar and Cay, 2006). Although, statistically non significant, there is a tendency for improvement in SOC and TN. The increase in SOC and TN in the less plowed soils could be due to the decreased mineralization rate of soil organic matter (Ozpinar and Cay, 2006). High temperatures in the study area (average maximum 31°C and minimum 15°C) could cause high oxidation of organic carbon (Clark and Gilmour, 1983).

Table 5.10. Effect of tillage systems on soil physical and chemical properties

Treatment	Total Nitrogen (%)	Organic Carbon (%)	Bulk density (gm cm^{-3})	pH in H$_2$O
CONV	0.074	0.62	1.35	8.18
STS	0.082	0.62	1.38	8.25
ST	0.079	0.64	1.39	8.23
	NS	NS	NS	NS

5.2.5 Financial analysis

The financial analysis results (Tables 5.11-5.13) showed that ST was superior to both CONV and STS. Conservation tillage systems save labor and traction needs for tillage although there are additional costs in weeding. However, the net result was in favor of the strip tillage system because weeding operation is cheaper (8 Birr-day^{-1}) than tillage operation (30 Birr-day^{-1}). STS had the least return because of the added cost of subsoiling and lesser grain yield when the time gap between tillage and planting was extended which resulted in delayed seedling emergence. However, with proper timing of the subsoiling operation, STS can be the most profitable tillage system considering the effect of disrupting the plow pan on infiltration and root growth. Moreover, local fabrication of the tool can reduce the price of the Subsoiler further improving its profitability.

5.2.6 Lessons from Participatory research

Involving farmers in research is a useful technique because they not only advise the researcher about important conditions specific to the area where a particular research is going to be undertaken, but they also follow the trials with curiosity. They tend to adopt research results more readily if they are involved in the research from the beginning. It is important to make farmers aware of the objectives and methodologies of the research so that they fully understand what is going on in their farms and can also protect the experimental plots.

Table 5.11 Cost of implements use

Operation	Price	Service life	Unit time		Cost	
	(Birr)	(hrs)	(hr-ha^{-1}-operation^{-1})		(Birr-ha-$^{-1}$-operation^{-1})	
			Melkawoba	Wulinchity	Melkawoba	Wulinchity
M[11]	25	200	22.67	27.1	2.83	3.39
R	25	200	13.5	15.6	1.69	1.95
S	30	200	9.6	11.8	1.44	1.77

At the beginning, the author faced some problems when working in some of the fields the owners of which were not fully acquainted with the research objectives and methodologies. Those farmers did not care even when the pegs were pulled out by children. Others thought that the researcher was a private investor who would take their lands and displace them. They were surprised to find out later that the researcher was only interested in numbers when they see their grains

---

[11] M means full tillage with *Maresha*, R means two passes with *Maresha* at 75 cm and S means single operation with the Subsoiler

returned to them after weighing. Once they were made aware of the activities taking place, not only they guarded the fields but they also tipped the researcher with some useful information about the behavior of the crops under the different treatments that were being evaluated. The farmers became more and more enthusiastic about the research and eager to know the outcomes. At the end of the research season more farmers were coming forward to try new techniques with the researcher.

Table 5.12(a) Financial analysis at Melkawoba (mean values)

Treatment	Costs: Birr-ha^{-1}				Total cost (Birr-ha^{-1})	Revenue (Birr-ha^{-1})	Net Benefit
	Tillage	Weeding	Implement				
CONV	383	450	3xM	9	842	2546	1705
STS	244	604	R+S	3	851	2482	1630
ST	182	627	R	2	811	2579	1768

Table 5.12 (b) Financial analysis at Melkawoba (Melkawoba 2005)

Treatment	Total cost (Birr-ha^{-1})	Grain yield (kg-ha^{-1})	Stock yield (kg-ha^{-1})	Revenue (Birr-ha^{-1})	Net Benefit
CONV	842	1847	2748	2996	2154
STS	851	2098	3788	3569	2718
ST	811	1954	4265	3509	2698

Table 5.13 Financial analysis at Wulinchity

Treatment	Costs (Birr-ha^{-1})		Implement use	cost	Total cost (Birr-ha^{-1})	Revenue	Net Benefit
	Tillage	Weeding					
CONV	437	490	3xM	10	937	2254	1317
STS	266	642	R+S	6	914	2173	1259
ST	201	574	R	5	781	2238	1458

Farmers normally undertake research in their own way. Over centuries, they have developed tillage systems that they think were optimum for the situation in which they operate. Some innovative farmers try new things either by chance or out of curiosity. If they succeed, they normally pass their experience to neighbors, friends and relatives. The other farmers also try the new technique and if it is found better than what they have been practicing, it gets accepted and gradually adopted by the community. Knowledge is then transferred verbally from parents to children. Migrating farmers, too, demonstrate their skills in areas where they settle and if what they introduce is important they earn recognition by the community. Farmers seldom divide their plots and compare different practices. But they compare one field with the other. Higher variations in field conditions and errors in measurements and reporting are compensated by large numbers of replications both in space and time.

Farmers in many cases believe that they have better understanding of their environment than extension agents and researchers. However, they are usually humble when they are approached. They try to please strangers by agreeing to what the strangers say. In most cases, elder farmers are more resistant to accepting new practices but once they are convinced they take things seriously. They are sources of knowledge and can influence the community easily about a new technology. Although, young farmers readily accept new practices and technologies they put less challenge to researchers who may be misled and draw the wrong conclusion that the technology they came with works perfectly in the area. Moreover, older farmers tend to ridicule technologies that are being adopted by the younger ones and this can cause a drawback to the adoption of new technologies. Hence, involvement of a combination of different age groups and gender can help accelerate participatory research, technology development and adoption.

5.3   Conclusions

The strip tillage system that involved subsoiling (STS) resulted in the least surface runoff, highest plant transpiration and highest grain yield followed by the strip tillage system without subsoiling (ST) when the days between the last tillage operations in STS/ST and planting (DTP) was 6 days. The reverse occurred when DTP was longer than 26 days.

A simple conceptual model simulated soil moisture better than a physically based complex model known as CoupModel, because preferential flows that are dominant over the root zone were not considered in the CoupModel. It is recommended that preferential flows be given due consideration when analyzing the hydrology of the unsaturated zone in semi-arid tropics. Closing furrows in STS/ST treatments gave significantly higher grain yield apparently because of reduced soil evaporation. Fertilization had a significant effect on grain yield of maize except in seasons when there was severe moisture stress problem. Water productivity for total evaporation and rainfall was the highest in STS showing efficient use of rainwater by the tested conservation tillage practices.

Tillage systems had no significant effect on soil physical and chemical properties after the three years period of the experiment. Financial analysis carried out on the average yields of the three years showed that ST was the most profitable tillage system while STS had the highest profitability when the time between the last tillage operation in STS/ST and planting was less than a week. It is recommended that additional studies be carried out in order to verify the effects of time of subsoiling on rainfall partitioning and yields of maize.

Involving farmers in research from trial initiation to implementation and follow up can improve performance in terms of better understanding of local problems, quicker development and adoption of technologies. Moreover, farmers become enthusiastic to try out new ideas and findings with greater possibilities of adapting technologies to local conditions.

Chapter 6

WATER PRODUCTIVITY OF IMPROVED AND MINIMUM TILLAGE SYSTEMS FOR *TEF* PRODUCTION IN SEMI-ARID ETHIOPIA[12]

6.1 Overview

In Ethiopia, *tef (Eragrostis tef* (Zucc) Trotter) (Figure 6.1) is the staple food for the majority of the population and covers 32% of the cultivated land (CSA, 1995). It is grown over a wide soil moisture regime because of its relative resistance to water logging in highland vertisols and to moisture stress in dry semi-arid areas (Ketema, 1997). *Tef* is endemic to Ethiopia and its major diversity is found only in that country. Although, the exact date and location for the domestication of *tef* is unknown, there is no doubt that it is a very ancient crop in Ethiopia where domestication took place more than 2000 years ago (Ketema, 1997).

Figure 6.1. *Eragrostis tef* (Zucc.) Trotter. (a) Inflorescence (b) branch of panicle with floret. (Source: Ketema, 1997)

---

[12] Based on: Temesgen, M., Rockström, J., Savenije, H. H. G., Hoogmoed, W. B. Water Productivity of Improved and Minimum Tillage Systems for *tef* Production in Semi-arid Ethiopia. Submitted to *Agricultural Water Management.*

Seedbed preparation for *tef* is often characterized by intensive tillage using the traditional *Maresha* plow (Figure 1.4). Seed bed preparation for *tef* requires 5 to 9 passes with the *Maresha* plow (Taddele *et al.*, 1993; Teklu and Gezahegn, 2003). In semi-arid areas where rainfall is low, farmers plow the land up to 5 times (see Chapter 3). Repeated tillage causes excessive pulverization leading to structural damage (Reicosky, 2001; Bezuayehu *et al.*, 2002); it requires too much traction and time making labor and oxen less productive and it also causes loss of soil through erosion. One of the reasons for repeated tillage is the presence of unplowed strips of land between adjacent V-shaped furrows created by the traditional *Maresha* plow (Figures 3.2 and 1.5) which also leaves clods that force farmers to carry out repeated tillage. Moreover, due to the need for cross plowing with the traditional tillage implement, one of any two consecutive tillage operations have to be laid along the slope, thereby encouraging surface runoff and soil erosion. Kruger *et al.*, (1996) estimated that high tillage frequency and other management problems have caused soil erosion, seriously affecting over 25% of the Ethiopian highland, while over 4% are irreversibly degraded.

Conservation tillage systems are believed to reduce surface runoff and maximize infiltration thereby making more water available to crop growth (Ahenkorah, *et al.*, 1995; Chen *et al.*, 1998; Steiner, 1998; Rockström and Jonsson, 1999). However, most of the conservation tillage systems developed so far involve row planting either in the form of direct planting (no tillage) or planting along cultivated lines (strip tillage). Since *tef* cannot be planted in rows, the conservation tillage systems developed for row planted crops cannot be directly applied thus calling for a different approach (IIRR and ACT, 2005).

Past experiments carried out on conservation tillage for *tef* production largely involved mere reduction of tillage frequency while using the same traditional tillage implement, the *Maresha* plow. A three-year trial conducted to study the effect of frequency of tillage on *tef* production at three locations in central Ethiopia showed that five times plowing with *Maresha* (the highest frequency) gave significantly higher grain yield compared to lower tillage frequencies (Tadele *et al.*, 1999). The authors associated high tillage frequency with the need for the preparation of fine seed bed and for controlling weeds. Reducing tillage frequency was also compensated by use of non-selective herbicide to control weeds. Erkossa *et al.*, (2006) conducted experiments on reduced tillage using herbicides for *tef* production in the highland vertisols reporting grain yield advantages of 8% over traditional systems. Extensive demonstrations were carried out in Ethiopia by the Sasakawa Global 2000 project in collaboration with the representatives of Monsanto using herbicides to control weeds (Gebre *et al.*, 2001). They reported promising conservation tillage systems but expressed concerns on the affordability of external inputs such as herbicides by resource poor smallholder farmers. Ofori (1993) concluded that the issue of affordability of herbicides is a major setback to the introduction of no-till system in smallholder farming system.

Moreover, in semi-arid areas where the land is normally bare during the dry season due to little biomass production and dry season feed shortages, surface runoff can be too high if the soil is left undisturbed. Much of the reported surface runoff reduction from no-tillage is associated with sufficient ground cover that protects the soil from rainfall impact thereby reducing compaction, increasing infiltration and retarding the movement of water by acting as a physical barrier. Therefore, where soil cover cannot be maintained, carefully designed tillage system has to be carried out to allow infiltration and to retard the overland movement of water.

Muliokela *et al.,* (2001) suggested that for resource-poor smallholder farmers in Africa, alternatives to the use of herbicides in conservation tillage such as the use of improved implements that are modified forms of existing tillage implements be sought for. In Ethiopia, a number of tillage implements have been developed as modifications to the traditional *Maresha* plow with a view to making them affordable, light and easy for use by smallholder farmers (See Chapter 4). An experiment was, therefore, initiated to test different types of tillage systems using the improved implements with the objective of selecting appropriate conservation tillage systems that can improve water productivity through increased infiltration, that can reduce surface runoff and that can reduce tillage frequency.

In this Chapter, the methodologies and results of a three year on-farm experiment carried out in selected semi-arid regions in Ethiopia on improved tillage systems for *tef* production are presented. The study assessed the possibilities to avoid cross plowing and to reduce surface runoff using conservation tillage systems specifically developed for *tef.* Moreover, possibilities of improving grain yield and water productivity of broadcast crops such as *tef* with the application of conservation tillage systems and the question of whether there will be any effect of different tillage systems on soil physical and chemical properties as well as the economics of the newly proposed tillage systems were addressed.

6.2 Materials and methods
6.2.1 Experimental site
The experiment was carried-out in 2003, 2004 and 2005 at Melkawoba and 2003 and 2004 at Wulinchity, which are typical dry semi-arid regions located in the central rift valley of Ethiopia (Figure 1.3). The experimental sites, Melkawoba and Wulinchity, are fully described in Chapter 3. The same farmers that were involved in the maize experiment were also involved in the *tef* experiments.

6.2.2 Treatments
Four different types of treatments have been tested:
1) Conventional tillage (CONV) involved three to four times plowing, which is the exact copy of farmers' practice with no limitations in oxen and labor.
2) Improved tillage with subsoiling (ITS) involved plowing once using the traditional tillage implement, *Maresha*, across the slope at the time when farmers undertake the first tillage. No tillage was performed until 3

weeks before planting during which the *Maresha* Modified Plow was used once perpendicular to the slope followed by subsoiling in the same direction at a spacing of 0.75 m. At planting, the sweep was used to control weeds and to incorporate Di-Amonium Phosphate (DAP) followed by broadcasting of seeds.

3) Minimum tillage (MT) involved making furrows perpendicular to the slope using the traditional tillage implement, *Maresha*, at a spacing of 0.75 m followed by subsoiling over the furrows. At planting, the sweep was used to control weeds followed by broadcasting of seeds.

4) Improved tillage (IT) involved making furrows perpendicular to the slope using the traditional tillage implement, *Maresha*, at the time when farmers undertake the first tillage. No tillage is performed until 3 weeks before planting during which the *Maresha* Modified Plow was used once perpendicular to the slope. At planting, the sweep was used to control weeds and to incorporate DAP fertilizer followed by broadcasting of seeds.

Seed rates were 30 kg-ha^{-1} with DAP applied at planting at a rate of 100 kg-ha^{-1} and Urea applied 35 days after planting at a rate of 50 kg-ha^{-1}.

6.2.3. Rainfall partitioning

Rainfall was measured daily at 9:00 am using two rain gauges installed near the experimental plots. Surface runoff was measured using rectangular troughs installed at the bottom of each plot (Figure 6.2). The size of the rectangular troughs was 10 m x 0.4 m x 0.2 m. The troughs were covered with plastic sheets supported on wooden frames that allow water from one side. Water collected in the trough was scooped daily at 9:00 am and measured using cans of predetermined capacity and graduated cylinders. Soil moisture was monitored using a Time-Domain Reflectometer (TDR) moisture measuring equipment from Eijkelkamp® and access tubes buried to depths of 1.8 m. Two access tubes were buried in each plot of treatments 1, 2 and 3 with two replications. Evaporation from a class A-pan was measured daily at 9:00 am which was used in a conceptual model developed to simulate rainfall partitioning.

6.2.4 Conceptual model to simulate rainfall partitioning

A conceptual threshold model was used to simulate rainfall partitioning at Wulinchity. The conceptual model (Figure 6.4) has been developed in the same way as described in Chapter 5, section 5.2. The difference is mainly the root depth. Thus, a root depth of 0.6 m has been assumed for *tef* as opposed to 1 m for maize. Figures 6.5-6.7 show measured and simulated soil moisture in *tef* over the root depth of 0.6 m at Wulinchity in 2004.

6.2.5 Grain yield

Five samples each from 2 m x 2 m area were harvested from each plot for biomass and grain yield assessment. The sampling areas were located 2 m from each side of the four corners while the fifth was located at the center of the plot. Biomass was measured using a stationary balance of 20 kg capacity in the field.

Threshing was carefully carried out by hand. Grain yield was measured using sensitive balances. The grain samples were dried in an oven at 70°C for 24 hours and the weight was adjusted to a moisture content of 15.5%. Water productivity was calculated following the methods described in Chapter 5. Statistical analysis was carried out using the SAS software (SAS Institute Inc., 1999).

6.2.6 Grain yield

Five samples each from 2 m x 2 m area were harvested from each plot for biomass and grain yield assessment. The sampling areas were located 2 m from each side of the four corners while the fifth was located at the center of the plot. Biomass was measured using a stationary balance of 20 kg capacity in the field. Threshing was carefully carried out by hand. Grain yield was measured using sensitive balances. The grain samples were dried in an oven at 70°C for 24 hours and the weight was adjusted to a moisture content of 15.5%. Water productivity was calculated following the methods described in Chapter 5. Statistical analysis was carried out using the SAS software (SAS Institute Inc., 1999).

6.2.7 Soil physical and chemical properties

The methods used for the determination of soil physical and chemical properties are the same as those described in Chapter 5.

6.2.8 Financial analysis

Economic evaluation of the different tillage systems was made using the costs of tillage and weeding in each operation, and the gains from grain and biomass yield calculated based on current market prices. Costs that were the same for all the tillage systems were not included. The costs of implement use were calculated based on the current prices of the implements that are on sale while estimates were given to those not yet in the market (Table 6.7). Service lives were estimated for each tillage implement after consulting farmers as to how long the implements lasted in their respective areas while the time required for each operation were obtained from field measurements. The time required to complete the different operations was used to calculate the respective cost of implement use.

The costs of operation were calculated based on labor hiring rates of 8 Birr per day (1USD=8.7 Birr) and oxen hiring rate of 30 Birr-day^{-1}. Financial analyses were made by using the sum of costs of implement use, tillage operation and weeding as total expense and sales from *tef* grain and straw as revenues. The price of *tef* grain was assumed to be 3 Birr-kg^{-1} while that of *tef* straw was assumed to be 0.3 Birr-kg^{-1}.

6.3 Results and discussion

6.3.1 Rainfall partitioning

Runoff: runoff (Q_S) was the highest (48 mm-season^{-1}) in the minimum tillage (MT) while it was the least (23 mm-season^{-1}) in the improved tillage system with subsoiling (ITS). Conventional tillage (CONV) resulted in a higher runoff (34 mm-season^{-1}) than ITS (Table 6.3).

Figure 6.2. Runoff collecting trough installed at the bottom of a 10 m x 10 m plot.

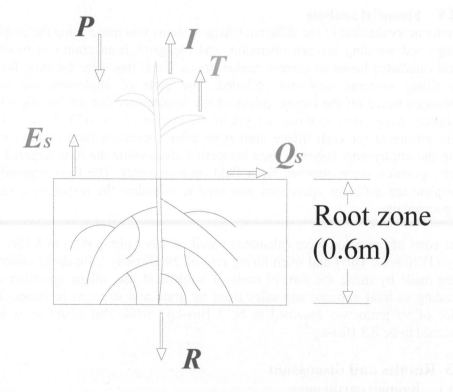

Figure 6.4. Conceptual model showing rainfall partitioning for *tef* crop. *P*: precipitation, Q_S : surface runoff, *I*: Interception, *T*: transpiration, E_S: Soil evaporation, *R*: Drainage

Table 6.1. Parameters used in the conceptual model

Property	Value	Unit
Crop coefficient, K_C	0.55	
Moisture content at field capacity, S_{FC}	19	%
Moisture content at wilting point, S_W	10	%
p	0.4	-
r	0.3	-
Interception threshold, I_D	4	mm-d^{-1}
Leaf overlap factor, C_C	0.95	-
Soil evaporation coefficient, K_S	0.5	-
Drainage coefficient, R_C	0.04	-

Table 6.2. Physical and chemical properties of soils at the initiation of the experiment at Melkawoba.

Soil texture	Sand	Silt	Clay
(Particle size distribution in %)	54	32	14
Bulk density (gm cm$^{-3)}$	1.35		
Organic carbon (%)	0.85		
Total Nitrogen (%)	0.10		

Regression analysis of data on rainfall and runoff yielded the relationships shown in Equations 6.1-6.3.

$$Q_{S(CONV)} = 0.12(P-6)$$
$$R^2 = 0.88$$

(6.1)

Where Q_S is surface runoff in mm-d^{-1} and P is daily rainfall in mm-d^{-1}.

$$Q_{S(ITS)} = 0.08(P-7)$$
$$R^2 = 0.9$$

(6.2)

$$Q_{S(MT)} = 0.16(P-5)$$
$$R^2 = 0.85$$

(6.3)

The figures in the bracket were used as thresholds to calculate the net rainfall, P_{NET}. The relationship between P_{NET} and Q_S is shown in Figure 6.3.

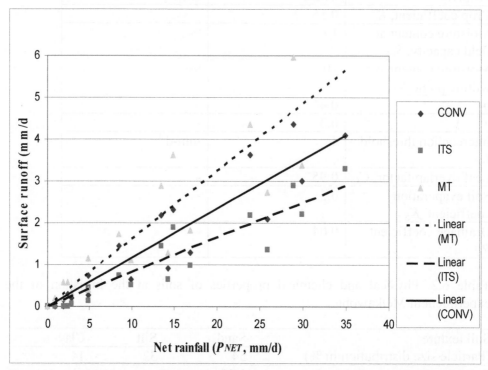

Figure 6.3. Surface runoff from *tef* fields as a function of net rainfall under different tillage systems. Net rainfall was obtained after subtracting runoff threshold for each treatment.

MT resulted in the highest surface runoff because there was no sufficient soil cover in the form of crop residues or cover crops and the soil was compacted by rainfall leading to lower rates of infiltration. Similar results were observed in other studies (Ajuwon, 1983; Roth *et al.,* 1988; Smith *et al.,* 1992; Bradford and Huang, 1994; Omer and Elamin, 1997). No-tillage has to be accompanied by soil cover that can protect the soil from the impacts of rainfall. Soil cover, either in the form of crop residues or cover crops grown during off-season, can also act as barriers to the movement of water thereby reducing runoff.

ITS resulted in the least runoff because:
1. Cross-plowing was avoided. Conventional tillage involves cross-plowing in order to disturb unplowed strips of land left by the traditional *Maresha* plow, which encourages surface runoff and erosion (Ndiaye *et al.,* 2005). The use of the traditional *Maresha* plow across the slope, during the first tillage, followed by plowing along the same direction with the Modified *Maresha* Plow (MMP), could be the main reason for reduced surface runoff in ITS.
2. Deeper penetration by MMP (Chapter 4) could have contributed to more infiltration.

3. Subsoiling carried out immediately after the use of MMP could have disrupted the plow pan (Chapter 4) resulting in increased infiltration (Sojka *et al.,* 1993).

Table 6.3. Effect of tillage systems on rainfall partitioning (mm-season$^{-1(13)}$) in *tef* at Wulinchity, 2004.

Treatment	P	Qs	I	T	Es	R	S	(T+R)/P	Qs/P	(I+Es)/P
CONV	431	34	101	49	126	228	-100	0.64	0.08	0.53
ITS	431	23	101	53	124	262	-124	0.73	0.05	0.52
MT	431	48	101	32	141	201	-86	0.54	0.11	0.56

P: precipitation, Q_S : surface runoff, I: Interception, T: transpiration, E_S: Soil evaporation, R: Drainage, ΔS: Change in water storage over the root zone.

Transpiration: Transpiration (T) was the highest in ITS apparently because more water was available in the root zone. Since transpiration is the useful component of rainfall, ITS, is the most preferred tillage system in making more water available to the crop.

Drainage: Drainage (R) was the highest in ITS. This could be because of increased infiltration. Higher drainage resulting from reduced surface runoff is preferable. In addition to conserving soil and nutrients that are washed away with surface runoff, evening of stream discharge by reducing flows during the rainy season and increasing the same later during the dry season can benefit downstream users. Besides, silting problems can be reduced. Table 6.3 also shows highest ratio of the useful components ($T + R$) to total rainfall (P) in ITS and the least ratio of surface runoff, Q_S, to P and the total unproductive evaporation ($I+E_S$) to P.

6.3.2 Water productivity

Results on water productivity are shown in Table 6.5. As in the case of maize (Chapter 5) water productivity for plant transpiration did not show appreciable differences among the tillage treatments. On the other hand, water productivity for total evaporation, W_{PET}, and water productivity for rainfall, W_{PP}, were the highest for ITS followed by CONV. MT gave the least values as grain yield was the lowest and surface runoff and unproductive evaporation were the highest. The effect of vapor shift as explained for maize in Chapter 5, Section 5.2.3 also applies for *tef.* The top 0.15 m layer of the soil under MT had the highest bulk density (Table 6.6) which could have increased surface runoff. The soil moisture storage was the highest for ITS followed by CONV and MT (Figure 6.8). This could be because of more infiltration in ITS following deeper plowing by the MMP and disruption of the plow pan by subsoiling.

---

[13] Season refers to period between seedling emergence and harvesting.

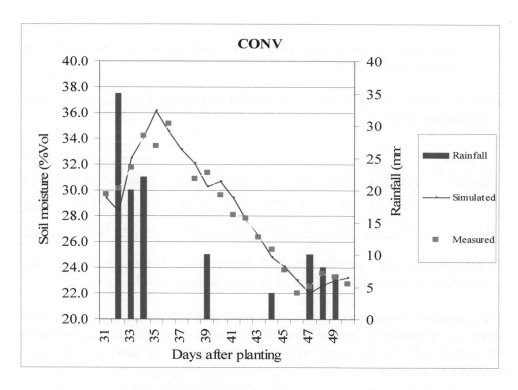

Figure 6.5. Simulated and observed soil moisture over the root zone (0-0.6m) in conventional tillage (CONV, Wulinchity, 2004)

The difference between ITS and IT is subsoiling. It is interesting to note that subsoiling was much more effective in *tef* than in maize (Chapter 5). Time of subsoiling could have an effect on the performance differences of the two tillage systems (ITS in *tef* and STS in maize). In the case of STS, most of the subsoiling was carried out long before planting time while in ITS, subsoiling was carried out few days before planting. Even in the case of STS, the best performance of subsoiling was obtained when there was a short gap (6 days) between subsoiling and planting.

6.3.3 Grain yield
The maximum grain yield was obtained from ITS (Table 6.4). This could be because more water was available to the crop in ITS. Moreover, deeper penetration by the MMP and by the Subsoiler (Chapter 4) could have resulted in deeper root growth. Better weed control by MMP could also have contributed to higher yields.

Outputs from the conceptual model
Figures 6.5.- 6.7 show model outputs for the days during which access tube readings were available. The simple conceptual model has reasonably estimated the soil moisture content over the root depth. Similar fits were obtained in the maize experiment (Chapter 5).

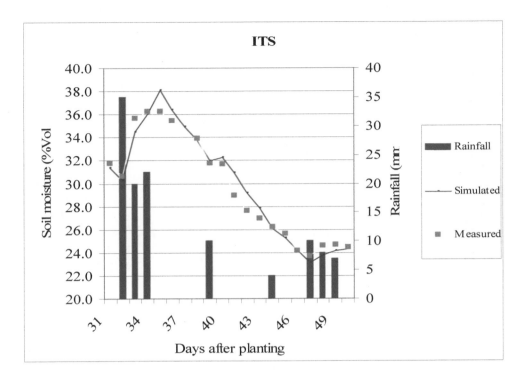

Figure 6.6. Root zone (0-0.6m) soil moisture simulated by the conceptual model versus observed data for improved tillage with subsoiling (ITS, Wulinchity, 2004)

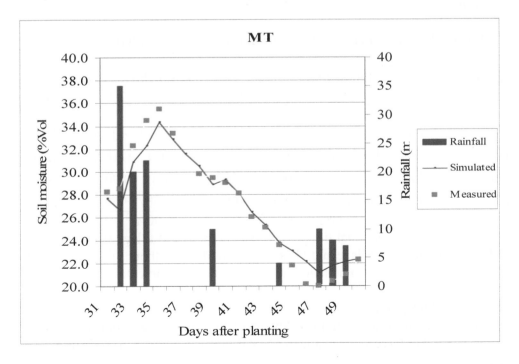

Figure 6.7. Root zone (0-0.6m) soil moisture simulated by the conceptual model versus observed data for minimum tillage (MT). (Wulinchity, 2004)

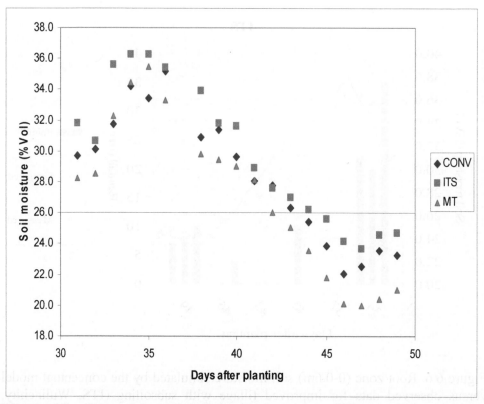

Figure 6.8. Comparison of soil moisture under different tillage systems (Wulinchity, 2004).

Table 6.4. Effect of tillage system on grain yield of *tef* (kg-ha)$^{-1}$.

Treatments	Melkawoba				Wulinchity		
	2003	2004	2005	**Mean**	2003	2004	**Mean**
CONV	900	960 (ab[14])	920(b)	**930**	1260(ab)	1070(ab)	**1170**
ITS	930	1160 (a)	1230(a)	**1110**	1460(a)	1180(a)	**1320**
MT	850	730 (b)	850(b)	**810**	1200(b)	890(b)	**1050**
IT	900	890 (b)	920(b)	**900**	1260(ab)	960(b)	**1110**
	NS	P>90%	P>90%		P>90%	P>90%	

CONV means conventional tillage, ITS means Improved tillage with subsoiling, MT means minimum tillage and IT is improved tillage without subsoiling.
Means followed by the same letter are not statistically different.

6.3.4 Soil properties

Tillage treatments did not significantly alter soil physical and chemical properties after a period of three years (Table 6.6). SOC and TN contents of soils take

---

[14] Means followed by the same letter are not statistically significant at P>90%

Table 6.5 Water productivity as affected by tillage systems in *tef* (kg grain m^{-3})

	Wulinchity 2004		
Treatment	W_{PT}	W_{PET}	W_{PP}
CONV	2.20	0.39	0.25
ITS	2.23	0.42	0.27
MT	2.28	0.32	0.21

longer (>5 years) to respond to reduced tillage (West and Post, 2002; Heenan *et al.*, 2004) while other investigators reported significant changes in shorter periods of two to three years (Su *et al.*, 2004; Ozpinar and Cay, 2006). The bulk density was the least in ITS and the highest in MT. The low bulk density observed in ITS could have permitted better infiltration and deeper root growth.

Table 6.6. Effect of tillage systems on soil physical and chemical properties

Treatment	Total Nitrogen (%)	Organic Carbon (%)	Bulk density (gm cm^{-3})	pH in H$_2$O
CONV	0.10	0.84	1.34 (ab)	8.3
ITS	0.10	0.84	1.30 (b)	8.2
MT	0.10	0.85	1.37(a)	8.2
	NS	NS	P>90%	NS

Means followed by the same letter are not statistically different.

6.3.5   Financial analysis

ITS resulted in the highest net benefit followed by CONV and IT both at Melkawoba and Wulinchity (Tables 6.8 and 6.9). MT gave the least net benefit. The net benefit has been mainly influenced by the amount of grain and straw yield and tillage costs. The new tillage systems increased weeding time but reductions in tillage time, which is more expensive than weeding (30 Birr-d^{-1} compared to 8 Birr-d^{-1}) favored ITS.

6.4 Conclusions

The Improved Tillage System (ITS) that involves the use of the *Maresha* plow across the slope followed by plowing once with the *Maresha* Modified Plow, subsoiling at a spacing of 0.75 m and planting with the Sweep, resulted in the least surface runoff, highest plant transpiration, highest grain and highest straw yield in *tef*. ITS was also the most profitable tillage system. On the other hand, MT resulted in the highest surface runoff, least plant transpiration, least grain and least straw yield. ITS is recommended for *tef* production in semi-arid environment under smallholder farming system. A simple conceptual threshold model reasonably estimated soil moisture content over the root depth.

Table 6.7. Cost of implements

Type of implement	Price (Birr[15])	Service life (hrs)	Unit time (hr-ha⁻¹-operation⁻¹)		Cost-ha⁻¹-operation⁻¹	
			Melkawoba	Wulinchity	Melkawoba	Wulinchity
Modified *Maresha* plow	250	500	27.20	31.6	13.60	15.80
Maresha (Full tillage)	25	200	22.67	27.10	2.83	3.39
Maresha at 75 cm	25	200	11.34	13.56	1.42	1.70
Maresha at 37.5 cm	25	200	13.50	15.6	1.69	1.95
Subsoiler	30	200	9.60	11.80	1.44	1.77
Sweep	40	200	9.83	10.20	1.97	2.04

Table 6.8 Financial analysis at Melkawoba

Treatment	Costs of operations and depreciation (Birr-ha⁻¹)				Revenue (Birr-ha⁻¹)[16]			Net Benefit
	Tillage	Weeding	Implements	Total	Grain	Straw	Total	
CONV	384	458	11	854	2790	1116	3906	3052
ITS	344	436	19	798	3330	1365	4695	3897
MT	135	584	5	724	2430	923	3353	2630
IT	297	470	17	785	2700	1080	3780	2995

Table 6.9 Financial analysis at Wulinchity

Treatment	Costs of operations and depreciation (Birr-ha⁻¹)				Revenue (Birr-ha⁻¹)			Net Benefit
	Tillage	Weeding	Implements	Total cost	Grain	Straw	Total	
CONV	558	488	14	1060	3510	1474	4984	3925
ITS	391	472	22	884	3960	1624	5584	4699
MT	153	656	6	815	3150	1103	4253	3438
IT	339	496	20	855	3330	1332	4662	3807

---

[15] 1 USD is equivalent to 8.7Birr.

[16] Price of *tef* grain is assumed to be 3 Birr-kg⁻¹ while that of *tef* straw is 0.3 Birr-kg⁻¹

Chapter 7

CONCLUSIONS AND RECOMMENDATIONS

7.1   Conclusions
7.1.1   The traditional tillage system.

Farmers in Ethiopia use an indigenous plow called *Maresha* for all stages of tillage. Because of the triangular geometry of the plow, V-shaped furrows are created that leave unplowed strips of land between adjacent passes. In order to deal with the unplowed strips of land, farmers carry out cross plowing. Cross plowing increases the time and energy requirement of seed bed preparation because the plow is moved over the already plowed area in order to access the unplowed parts. Moreover, cross plowing entails laying furrows along the hillslope in one of any two consecutive tillage operations, which can encourage high surface runoff.

Farmers start tillage as early as February and continue doing so until July. In most cases, *tef* fields are plowed 3 to 5 times while maize fields are plowed 3 to 4 times. Farmers realize that repeated plowing with the *Maresha* Plow causes evaporation losses due to exposure of the lower moist layers. However, they can not avoid plowing which is needed to break the surface crust formed by rainfall that follows a wetting-drying cycle. Moreover, the need for controlling weeds that emerge before planting and that cause loss of soil moisture through unproductive transpiration, forces farmers to plow during dry spells.

Farmers do not plow their fields before the rains start to avoid high draft power requirement, excessive pulverization leading to compaction by subsequent rains, higher weed infestation and formation of too many clods. They also want to let weeds emerge for a better control. The main purposes of tillage in the production of maize and *tef* at Melkawoba and Wulinchity are soil moisture conservation and weed control. Farmers also perceive soil warming as one of the purposes of tillage. Plow pans were found at depths ranging from 0.18 m to 0.25 m.

7.1.2   The *Maresha* modified conservation tillage implements

The implements that were developed as modifications to the traditional tillage implement, the *Maresha* Plow, were found to be suitable to the respective operations they were developed for while maintaining simplicity, light weight and low cost nature of the traditional plow. The Subsoiler, when operated along furrows made by the *Maresha* plow, penetrated up to a depth that enables disruption of the hard pan created under the traditional cultivation system. The use of the Row Planter to plant maize resulted in early and twice as much seedling emergence as manual placement of seeds, under moisture stress situations leading to increased grain yields, in addition to saving labor and time by up to 85%. The Tie-ridger made furrows with larger cross sectional areas than those made by the *Maresha* Plow and the inverted BBM while requiring lower draft forces. The lifting force required by the Tie-ridger, when tying furrows, was lower than that required by the *Maresha* plow and the inverted BBM. The *Maresha* Modified Plow reduced tillage frequency because of U-shaped furrow cross-sections and better weed control. It also increased depth of tillage leading to more infiltration and higher yields.

7.1.3   The locally adapted conservation tillage systems.

Reduced water productivity and hence lower crop yields in Ethiopia are caused by the traditional tillage system, which limits the water available to the crop. Among the conservation tillage systems tested on *tef* the one that involves initial soil opening with the *Maresha* Plow along the contour leaving narrow unplowed strips at a spacing of 0.75 m followed by one time plowing with the *Maresha* Modified Plow, subsoiling and planting with the Sweep (ITS) resulted in the least surface runoff, highest transpiration and the highest yields. Water productivity using total evaporation and rainfall were the highest for ITS followed by conventional tillage (CONV) and minimum tillage (MT). Minimum tillage performed worse than conventional tillage because of lower infiltration and higher weed infestation.

Among the conservation tillage treatments tested on maize, the one that involved strip tillage at 0.75 m spacing followed by subsoiling and planting over the same lines (STS) resulted in the least surface runoff, highest transpiration and the highest crop yields followed by the one that did not involve subsoiling (ST) and the traditional tillage system (CONV). However, when the time between the last tillage operation and planting was more than 26 days the reverse occurred because of higher weed transpiration and surface compaction by rainfall. The effect was more pronounced with an increase in the cumulative rainfall occurring between the last tillage operations in the STS/ST treatments and planting.

A simple conceptual model simulated soil moisture in the root zone better than a physically based model that employed Richards equations. This could be because preferential flows are dominant in the semi-arid tropics while the tested physically based model did not estimate the influences of such flows on the hydrology of the unsaturated zone.

Closing furrows in STS/ST treatments gave significantly higher grain yield apparently because of reduced soil evaporation. Fertilization had a significant effect on grain yield of maize except in seasons when there was severe moisture stress. Financial analysis carried out on the average yields of the three years showed that ST was the most profitable tillage system while STS had the highest profitability when the time between the last tillage operation in STS/ST and planting was less than a week. Tillage systems did not result in any significant difference in the physical and chemical properties of the soil.

The experiments have shown that it is indeed possible to introduce new insights and new technology into traditional farming systems, provided these innovations increase yields, reduce labor and are affordable. Farmers are eager to adopt a new technology if these conditions are met and if they are proven to be effective in the field. With the suggested additional research required to fill some of the gaps observed during the study, it is concluded that the tested conservation tillage systems can be applied to increase water productivity and grain production by positively altering rainfall partitioning in the dry semi-arid areas of Ethiopia while being affordable by smallholder farmers.

7.1.4   Participation of farmers in research.
Involvement of farmers beginning from the inception of on-farm research benefits both the researcher and the farming community. In addition to tapping their indigenous knowledge and properly managing research fields, involvement of farmers in research boosts their confidence in newly introduced technologies making them enthusiastic to try them by themselves. This has the potential of quicker development and adoption of appropriate technologies.

7.2   Recommendations
Based on the consistent performance of the improved tillage system tested on *tef*, ITS is recommended for popularization among smallholder farmers in semi arid areas. The *Maresha* modified implements can also be popularized among farmers.

The conservation tillage system that was tested on maize and that involved strip tillage followed by subsoiling (STS) has shown promising results when the time between the last tillage operation and planting was less than a week. Additional studies are required to verify the performance of the tillage system paying particular attention to the timing of subsoiling. The use of the sweep that does not expose the lower moist soil while breaking surface crust and controlling weeds during dry spells should be tested further as a means to solve the dilemma that farmers face during the dry extended periods between tillage commencement and planting.

Farmers believe that plowing helps to warm up the soil thereby improving seed germination. Further investigation is required to study the effect of plowing on soil temperature and thus on seedling emergence. Farmers' techniques of tillage timing, which they believe helps improve soil workability and infiltration and reduce soil evaporation should be explored further for possible incorporation in the design of appropriate conservation tillage systems.

Studies are required to measure the water balance at a watershed scale, which will enable the measurement of stream flows (surface and rapid ground water flows) that could have an impact on the uniformity of stream flows during rainy and dry periods, through increased infiltration and groundwater recharge, for the benefit of downstream users. Moreover, soil loss studies resulting from the application of conservation and conventional tillage systems should be carried out both on field scale and over watersheds. Further studies are required on preferential flows occurring in semi arid areas of Ethiopia for a better understanding of the hydrology of unsaturated flow.

Tied ridges described in Chapter 4 have been tested for many years with the objectives of improving soil moisture through the reduction of surface runoff with mixed results. This has limited the adoption of tied ridges in dry areas. However, this study has given an insight into the problem of soil evaporation resulting from dry spells occurring before planting. Therefore, trials should be carried out on tied ridges by comparing the operation before planting with one carried out after planting. The analogy between the findings with timing of subsoiling can be applied to tie ridging and possibly disclose the mysteries of mixed results reported on tie ridging. Moreover, lack of maneuverability of tie ridgers has been a set back to the adoption of the practice. This study has demonstrated the availability of a simple and easy-to-use implement for making tied ridges. Therefore, it is recommended that trials be initiated on tie ridging paying particular attention to the timing of the operation in relation to loss of soil moisture through evaporation.

Since farmers know some of the unique situations on their farms better than researchers, smallholder system innovations under local conditions should be investigated together with farmers. The potential of making research partnership with farmers to achieve quicker development and adoption of appropriate technologies should be given utmost importance.

REFERENCES

Addiscott, T.M., Dexter, A.R., 1994. Tillage and crop residue management effects on losses of chemicals from soils. Soil Till. Res. 30, 125–168.

Agishi, E.C. (1985). Forage Legumes and Pasture Development in Nigeria. In: Nuru, S. and Ryan, J.G. 1985 (eds). Proceedings of Nigeria / Australia Seminar on Collaborative Agricultural Research. Nov. 14-15, 1983. Shika, Nigeria. Pp 79-87.

Ahenkorah, Y.E., Owusu-Bennoah, G.N.N, Dowuona (eds). (1995). Sustaining Soil Productivity in Intensive African Agriculture. Published by the CTA. Postbus 380, 6700 AJ Wageningen, The Netherlands. 42p.

Ajuwon, S.O., 1983. Intrinsic deficiencies of zero-tillage. In: Proceedings of the First National Tillage Symposium, Nigeria Society of Agricultural Engineers, Ilorin, Nigeria, November 22–25, 25 pp.

Akinyemi, J.O., Akinpelu, O.E., Olaleye, A. O. 2003. Performance of cowpea under three tillage systems on an Oxic Paleustalf in southwestern Nigeria. Soil and Till. Res. 72, 75–83.

Akobundu, I.O. (1984). Advances in Live Mulch Crop Production in the Tropics. Proceedings of Western Society of Weed Science 37: 51-56.

Alvenäs, G., Jansson, P-E., 1997. Model for evaporation, moisture and temperature of bare soil: calibration and sensitivity analysis. Agr. Forest Meteorol., 88: 47-56.

Antunes, M.A.H., Walter-Shea, E.A., Mesarch, A.M. 2001. Test of an extended mathematical approach to calculate maize leaf area index and leaf angle distribution. Agricultural and Forest Meteorology. 108, 45–53.

Basic, F., Kisic, I., Butorac, A., Nestroy, O., and Mesic, M. (2001). runoff and Soil Loss Under Different Tillage Methods on Stagnic Luvisols in Central Croatia. *Soil Till. Res. 62, 145-151.*

Bauer, A., Black, A.L., 1994. Quantification of the effect of soil organic matter content on soil productivity. Soil Sci. Soc. Am. J. 58, 185–193.

Baumhardt, R.L., and Jones, O.R. 2002. Residue management and tillage effects on soil-water storage and grain yield of dryland wheat and sorghum for a clay loam in Texas. Soil Till Res. 68, 71–82

Bayu, W., Alemu, G., K/Mariam, G., Ewnetu, Dilnesa. (1998). BBM: Wings beyond what they were meant for. *AgriTopia* 13(2); April-June 1998.

Benites, J.R., Ashburner, J.E. (2001). FAO's role in promoting conservation agriculture. In: 1 World Congress on Conservation Agriculture, Madrid, 1-5 October, 2001. XUL Avda. Medina Azahara 49, pasaje, 14005 Cordoba, Spain. Pp 133-147.

Beyene, H., Negassa, A., Dadi, L., and Mulatu, T., 1990. Crop Production and agricultural Implements in the Bako, Holetta and Nazret areas. Research Report No. 11. Institute of Agricultural Research. P. O. Box 2003. Addis Abeba.

Bezuayehu, T., Gezahegn, A., Yigezu, A., Jabbar, M.A., Paulos, D. (2002) Nature and causes of land degradation in the Oromiya Region: A review. Socio-economics and Policy Research Working Paper 36. ILRI (International Livestock Research Institute), Nairobi, Kenya. 34p.

Biamah, E.K., Rockström, J. (2000). Development of sustainable conservation tillage systems. In: Biamah, E.K., Rockström, J., Okwach, G.E. (eds). Conservation Tillage for Dryland Farming. Technological Options and Experiences in Eastern and Southern Africa. Nairobi: Regional Land Management Unit, Swedish International Development Agency (Sida), 2000. RELMA Workshop Report Series 3. Pp 36-41.

Biamah, E. K., Gichuki, F. N., Kaumbutho, P. G. 1993. Tillage methods and soil and water conservationin eastern Africa. Soil and Tillage Research, 27, 105-123

Blombäck, K., Stähli, M., Eckersten, H., 1995. Simulation of water and nitrogen flows and plant growth for a winter wheat stand in Central Germany. Ecol. Model., 81: 157-167.

Boa-Ampongsem, K., Bayorbor, T.B., Findley, J.B.R. 2001. Introduction and adoption of no-till crop production system form small-scale farmers in Ghana. In: 1 World Congress on Conservation Agriculture, Madrid, 1-5 October, 2001. XUL Avda. Medina Azahara 49, pasaje, 14005 Cordoba, Spain. Vol. II, Pp 67-71.

Bradford, J. M., Huang, C. 1994. Inter-ril soil erosion as affected by tillage and residue cover. Soil and Tillage Research 31, 353-361.

Bruneau, P. M. C., Twomlow, S. J. 1999. Hydrological and Physical Responses of a Semi-arid Sandy Soil to Tillage. *J. Agric. Engng. Res.,* 72, 385-391.

Castrignano, A., Maiorano, M., Prudenzano, M., Fornaro, F., Cafarelli, B. (2001). Assessment of the Effect of Tillage and Crop Residue Management on Soil Impedance Taking into Account Spatial Heterogeneity. In: 1 World Congress on Conservation Agriculture, Madrid, 1-5 October, 2001. XUL Avda. Medina Azahara 49, pasaje, 14005 Cordoba, Spain. Vol. II, Pp 237-241.

Chen, J., HongWen, W., Li, H.W. (1998). Technology and Machinery System of Mechanized Conservation Tillage for Dry Land maize. Journal of China Agricultural University. 3:4. 33-38.

Clark, M.D., Gilmour, J.T. 1983. The effect of temperature on decomposition at optimum and saturated soil water contents. Soil Sci. Soc. Am. J. 47, 927–929.

CSA (Central Statistics Authority). 1995. Agricultural Sample Survey 1994/95. Report on Area and Production for Major Crops. Vol:1. CSA, Addis Abeba.

Day, J. C., Hughes, D.W, Butcher, W. R. 1992. Soil, water and crop management alternatives in rainfed agriculture in the Sahel: an economic analysis. *Agricultural Economics,* 7, 267-287.

Derpsch, R. (1998). Historical review of no tillage cultivation of crops. In: Proceedings of the First JICAS Seminar on Soybean Research. No-tillage cultivation and future research needs, March 5-6, 1998, Iguassu, Brazil, JIRCAS Working Report No. 13. Pp 1-18

Derpsch, R., Moriya, K. (1998). Implications of no-tillage versus soil preparation on sustainability of agricultural production. Sustainable Land Use – Furthering Cooperation Between People and Institutions, Advances in Geoecology 31, Vol. II, Catena Verlag, Reiskirchen, 1998, p 1179-1186.

Diaz-Zorita, M., Duarte, G. A., Grove, J. H. (2002). A Review of No-till Systems and Soil Management for Sustainable Crop Production in the Subhumid and Semiarid Pampas of Argentina. Soil Till. Res. 65, 1-18.

Duley, F.L., Russel, J.C., 1939. The use of crop residues for soil and moisture conservation. J. Am. Soc. Agron. 31, 703–709.

Dumanski, J., R. Peiretti, J. Benetis, D. McGarry, and C. Pieri. 2006. The paradigm of conservation tillage. Proc. World Assoc. Soil and Water Conserv., P1: 58-64.

ECAF. (1999). Conservation Agriculture in Europe: Environmental, Economic and EU Policy Perspectives. European Conservation Agriculture Federation. Rond Point Schuman 6, box 5, B1040 Brussels, Belgium. 37p.

Edwards, W.M., Triplett, G.B., Van Doren, D.M., Owens, L.B., Redmond, C.E., Dick, W.A., 1993 Tillage Studies with a Corn-Soybean Rotation: Hydrology and Sediment Loss. Soil Sci. Soc. Am. J. 57, 1051-1055.

Elwell, H., Brunner, E., and Mariki, W. (2000). Recommendations on the Adoption of Minimun and Conservation Tillage in Tanzania. In: Conservation Tillage for Dry land Farming. RELMA/Sida, ICRAF House, Gigiri P.O.Box 63403, Nairobi, Kenya. 94p.

Engida, M. (2000). A Desertification Convention Based on Moisture zones of Ethiopia. Ethiopian Journal of Natural Resources. 2000 (1): 1-9.

Erkossa, T., Stahr, K., Gaiser, T.2006. Soil tillage and crop productivity on a Vertisol in Ethiopian highlands. Soil and Till. Res. 85, 200–211.

Fohrer, N., Berkenhagen, J., Heckler, J.M., Rudolph, A., 1999. Changing soil and surface conditions during rainfall—single rainstorm/subsequent rainstorms. Catena 355–375.

Freitas, V.H. (2000). Soil Management and Conservation for Small Farms. Strategies and Methods of Introduction, Technologies and Equipment. FAO Soils Bulletin 77. 31p.

Fuentes, J.P., Flury, M., Huggins, D. R., Bezdicek, D.F. 2003. Soil water and nitrogen dynamics in dryland cropping systems of Washington State, USA. Soil Till Res. 71 33–47

Gärdenäs, A.I., Jansson, P-E., 1995. Simulated water balance of Scots pine stands for different climate change scenarios. J. Hydrol., 166: 107-125.

GART yearbook (2001). Tillage Systems Comparison Trials. GART yearbook, Vol.3 (1). Golden Valley Agricultural Trust. P. O. Box 50834, Lusaka, Zambia. Pp. 36-38.

Gebre, T., Retu, B., Bekele, K., Dubale, P., Findlay, J.B.R. (2001). The introduction of conservation tillage system to small-scale farmers in Ethiopia. In: 1 World Congress on Conservation Agriculture, Madrid, 1-5 October, 2001. XUL Avda. Medina Azahara 49, pasaje, 14005 Cordoba, Spain. Pp765-768.

Georgis, K., Sinebo, W. (1993). Tillage, Soil and Water Conservation Research on maize in Ethiopia. In: Benti Tolessa and Joel K. Ransom (eds). 1993. Proceedings of the First National maize Workshop, May 5-7 1992, Addis Ababa, Ethiopia. pp56-61

Gicheru, P.T. 1994. Effects of residue mulch and tillage on soil moisture conservation. Soil Technology 7, 209-220.

Gill, K.S., Arshad, M.A., Chiunda, B.K., Phiri, B., Gumba, M. (1992) Influence of residue mulch, tillage and cultural practices on weed mass and corn yield grown in three field experiments. Soil Till. Res. 23, 211-223.

Gjettermann, B., Nielsen, K. L., Petersen, C. T., Jensen, H. E., Hansen, S. 1997. Preferential flow in sandy loam soils as affected by irrigation intensity. Soil Technol. 11, 139-152.

Goe, M. R. (1987). Animal traction on smallholder farms in the Ethiopian highlands. PhD Thesis, Cornell University. 211p.

Greene H. W., 2003. Econometric analysis. Fifth edition. Pearson Education Inc. New Jersey.

Gustafsson, D., Lewan, E., Jansson, P-E., 2004. Modeling water and heat balance of the Boreal landscape - comparison of forest and arable land in Scandinavia. J. Appl. Meteorol. 43: 1750-1767.

Guzha, A.C. 2004. Effects of tillage on soil microrelief, surface depression storage and soil water storage . Soil and Till. Res. 76, 105–114.

Heenan, D.P., Chan, K.Y., Knight, P.G. 2004. Long-term impact of rotation, tillage and stubble management on the loss of soil organic carbon and nitrogen from a Chromic Luvisol. Soil and Tillage Research 76, 59–68.

Hoogmoed, W.B. (1999). Tillage for soil and water conservation in the semi-arid tropics. Doctoral Thesis. Wageningen University, The Netherlands. 184p.

Hoogmoed, W. B., Stevens, P., Muliokela, S. W. 2003. Conservation tillage with animal draught in Zambia. In: Proceedings of ISTRO 2003, International Soil Tillage Research Organization Conference, Brisbane, Australia, pp 548-553.

Hoogmoed, W. B., Stevens, P., Samazaka, D., Buijsse, M. 2004. Animal Draft Ripping: The Introduction of a Conservation Farming Technology in Zambia. In: 2004 CIGR Int. Conference, (W Zhicai and G Huangwen, Eds.) China Agricultural Science and Technology Press Beijing, China pp 393-399.

Hulugalle, N.R. 1990. Water Conservation Alleviation of soil constraints to crop growth in the upland Alfisols and associated soil groups of the West African Sudan savannah by tied ridges. *Soil and Tillage Research,* 18, 231-247.

IGAD and FAO, 1995. Crop Production System Zones of the IGAD sub-region. Agro meteorology Working Paper Series No. 10. FAO, Rome, Italy.

IIRR and ACT. 2005. Conservation agriculture: A manual for farmers and extension workers in Africa. International Institute of Rural Reconstruction, Nairobi; African Conservation Tillage Network, Harare. ISBN 9966-9705-9-2. pp52.

Impens, I. and Lemeur, R., 1969: Extinction of net radiation in different crop canopies. Arch. Geoph. Bioklimatol., Ser. B., 17: 403-412.

ISSS-ISRIC-FAO, 1998. World Reference Base for Soil Resources. World Soil Resources Report 84. Rome.

Jaiyeoba, I.A. 2003. Changes in soil properties due to continuous cultivation in Nigerian semiarid Savannah. Soil and Till. Res. 70, 91–98.

Jansson, P.-E. and Halldin, S. 1979. Model for the annual water and energy flow in a layered soil. In: Comparison of forest and energy exchange models. Halldin, S. (ed.). Society for Ecological Modelling, Copenhagen. pp. 145-163.

Jansson, P.-E. and Karlberg, L., 2004. Coupled heat and mass transfer model for soil-plant-atmosphere systems. Royal Institute of Technology, Dept of Civil and Environmental Engineering., Stockholm, Sweden, 435 pp.

Jonsson, L.O., Singisha, M.A., Mbise, S.M.E. (2000). Dry land farming in Tanzania: Experiences from the Land Management Program. In: Biamah, E.K., Rockström, J., Okwach, G.E. (eds). Conservation Tillage for Dryland Farming. Technological Options and Experiences in Eastern and Southern Africa. Nairobi: Regional Land Management Unit, Swedish International Development Agency (Sida), 2000. RELMA Workshop Report Series 3. Pp 96-113.

Kamau, G.M., Ransom, J.K., Saha, H.M. (1999). maize cowpea rotation for weed management and improvement of soil fertility on sandy soil in coastal Kenya. In: Proceedings of the Sixth Eastern and Southern Africa maize Conference. 21-25 September, 1998. Pp 223-225.

Kannegieter, A. (1967). Zero cultivation and other methods of reclaiming Pueraria fallowed land for food crop cultivation in the forest zone of Ghana. Trop. Agrculturalist Vol. CXXV : 33-41.

Kaoma-Sprenkels, C., Stevens, P.A., Wanders, A.A. (2000). IMAG-DLO and conservation tillage: Activities and experiences in Zambia. In: Biamah, E.K., Rockström, J., Okwach, G.E. (eds). Conservation Tillage for Dryland Farming. Technological Options and Experiences in Eastern and Southern Africa. Nairobi: Regional Land Management Unit, Swedish International Development Agency (Sida), 2000. RELMA Workshop Report Series 3. Pp 42-53.

Karlberg, L., Ben-Gal, A., Jansson, P.-E. and Shani, U. 2005. Modelling transpiration and growth in salinity-stressed tomato under different climatic conditions. Ecol. Model. 190(1-2), 15-40.

Kaumbutho, P.G. (2000). Overview of conservation tillage practuices in Eastern and Southern Africa In: Biamah, E.K., Rockström, J., Okwach, G.E. (eds). Conservation Tillage for Dryland Farming. Technological Options and Experiences in Eastern and Southern Africa. Nairobi: Regional Land Management Unit, Swedish International Development Agency (Sida), 2000. RELMA Workshop Report Series 3. Pp1-26.

Kaumbutho, P.G., Simalenga, T.E. (eds). (1999). Conservation tillage with animal traction. A resources book of Animal Traction Network for Eastern and Southern Africa (ATNESA), Harare, Zimbabwe.134p.

Ketema, S. 1997. *tef. Eragrostis tef* (Zucc.) Trotter. Promoting the conservation and use of underutilized and neglected crops. 12. Institute of Plant Genetics and Crop Plant Research, Gatersleben/International Plant Genetic Resources Institute, Rome, Italy.

Kijne, J. W., Barker, R., Molden, D. 2003. Water Productivity in Agriculture: Limits and Opportunities for Improvement. CABI Publishing, Wallingford.

Kossila, V., 1988. The availability of crop residues in developing countries in relation to livestock populations. P 29 – 39. In: Reed J D, Capper B S and Neate P J (Eds) (1988). Plant breeding and the nutritive value of crop residues. Proceedings of a workshop held at ILCA, Addis Ababa, Ethiopia, 7 – 10 December 1987. ILCA, Addis Ababa.

Kruger, H., Berhanu, F., Yohannes, G., Kefeni, K., 1996. Creating an inventory of indigenous soil and water conservation measures in Ethiopia. In: Chris, R., Ian, S., Camilla, T. (Eds.), Sustaining the Soil Indigenous Soil and Water Conservation in Africa International Institute for Environment and Development, Earthscan, London.

Lal, R. (1983). No-till farming: Soil and water conservation and management in the humid and sub humid tropics. IITA monograph No. 2. 64p.

Lal, R. 2001. World cropland soils as a source or sink for atmospheric carbon. Adv. Agron. 71, 145–191.

Larney, F.J., Lindwall, C.W. 1995. Rotation and tillage effects on available soil water for winter wheat in a semi-arid environment. Soil & Tillage Research 36, 111-127.

Laryea, K.B., Pathak, P., Klaij, M.C. 1991. Tillage systems and soils in the semi-arid Tropics. Soil and Till. Res. 20, 201-218.

Lee, K. S., Park, S. H., Park, W. Y. 2003. Strip tillage characteristics of rotary tiller blades for use in a dryland direct rice seeder Soil & Tillage Research 71, 25–32.

Lemenih, M., Itanna, F. 2004. Soil carbon stocks and turnovers in various vegetation types and arable lands along an elevation gradient in southern Ethiopia. Geoderma 123, 177–188.

Lemenih, M., Karltun, E., Olsson, M. 2005. Assessing soil chemical and physical property responses to deforestation and subsequent cultivation in smallholders farming system in Ethiopia. Agriculture, Ecosystems and Environment. 105, 373–386.

Lenssen, A.W., Johnson, G.D., Carlson, .R. 2007. Cropping sequence and tillage system influences annual crop production and water use in semiarid Montana, USA. Field Crops Research 100, 2–43.

Licht, M.A., Al-Kaisi, M. 2005. Strip-tillage effect on seedbed soil temperature and other soil physical properties. Soil and Till. Res. 80, 233–249.

Lindroth, A. 1985. Canopy conductance of coniferous forests related to climate. Water Resour. Res. 21, 297-304.

Lohammar, T., Larsson, S., Linder, S. and Falk, S.O., 1980. FAST – simulation models of gaseous exchange in Scots pine. In: T. Persson (Editor), Structure and Function of Northern Coniferous Forests – An Ecosystem Study, Ecol. Bull., Stockholm, 32: 505-523.

MacRae, R.J., Mehuys, G.R. (1985). The effect of green Manuring on the physical properties of temperate soils. Advances in Soil Science 3: 71-94.

Mapa, R.B., Green, R.E., Santo, L., 1986. Temporal variability of soil hydraulic properties with wetting and drying subsequent to tillage. Soil Sci. Soc. Am. J. 50, 1133–1138.

Martinez-Raya, A., Fran, J.R., Ruiz-Gutierrez, S., Martinez-Vilela, A., Aguilar, J. (2001). Evaluation of Soil Protection with Different Types of Plant Cover. In: 1 World Congress on Conservation Agriculture, Madrid, 1-5 October, 2001. XUL Avda. Medina Azahara 49, pasaje, 14005 Cordoba, Spain. Pp431-434.

McGlynn, B.L., McDonnell, J., Brammer, D.D. 2002. A review of the evolving perceptual model of hillslope flowpaths at the Miami catchments, New Zealand. Journal of Hydrology. 257, 1-26.

Middleton, N., Thomas, D., (eds.). (1997). World Atlas of Desertification, 2nd Edition, UNEP. Arnold, London, UK. 181p.

Monjardino, M., Pannell, D.J. and Powles, S. (2000). The Value of Green Manuring in the Integrated Management of Herbicide-Resistant Annual Ryegrass (Lolium rigidum). (SEA Working Paper 00/11). 24p.

Monteith, J.L. 1965. Evaporation and the atmosphere. In: The state and movement of water in living organisms. 19th Symp. Fogg, G.E. (ed.). Soc. Exp. Biol., The Company of biologists, Cambridge. pp 205-234.

Motavalli, P.P., Stevens, W.E., Hartwig, G. 2003. Remediation of subsoil compaction and compaction effects on corn N availability by deep tillage and application of poultry manure in a sandy-textured soil. Soil and Tillage Research 71, 121–131.

Mulatu, T., Regassa, T., 1986. Nazret Mixed Farming Systems zone survey report. Research Report No, 2/87. Department of Agricultural Economics and Farming Systems Research, Institute of Agricultural Research, Addis Ababa.

Muliokela, S., Hoogmoed, W., Stevens, P., Dibbits, H. 2001. Consraints and possibilities for conservation farming in Zambia. In: World Congress on Conservation Agriculture, Madrid, 1-5 October, 2001. XUL Avda. Medina Azahara 49, pasaje, 14005 Cordoba, Spain. pp 61-66.

Mullins, G.L., Alley, S.E., Reeves, D.W. 1998. Tropical maize response to nitrogen and starter fertilizer under strip and conventional tillage systems in southern Alabama. Soil & Tillage Research 45, 1998.1–15

Ndiaye, B., Esteves, M., Vandervaere, J.P., Lapetite, J. M., Vauclin, M. 2005. Effect of rainfall and tillage direction on the evolution of surface crusts, soil hydraulic properties and runoff generationfor a sandy loam soil. Journal of Hydrology 307, 294–311.

Nitzsche, O., Kruck, S. T., Schmidt, W., Richter, W. (2001). Reducing Soil Erosion and Phosphate Losses and Improving Soil Biological Activity Through Conservation Tillage Systems. In: 1 World Congress on Conservation Agriculture, Madrid, 1-5 October, 2001. XUL Avda. Medina Azahara 49, pasaje, 14005 Cordoba, Spain. Pp179-184.

Nyagumbo, I. (2000). Conservation technologies for smallholder farmers in Zimbabwe. In: Biamah, E.K., Rockström, J., Okwach, G.E. (eds). Conservation Tillage for Dryland Farming. Technological Options and Experiences in Eastern and Southern Africa. Nairobi: Regional Land Management Unit, Swedish International Development Agency (Sida), 2000. RELMA Workshop Report Series 3. Pp 68-75.

Ofori, C.S. 1993. Towards the development and technology transfer of soil management practices for increased for agricultural production in Africa. In: Ahenkorah, Y., Owusu-Bennoah, E., Dowuona, G.N.N. (eds). Sustaining soil productivity in intensive African agriculture. Technical Centre for

Agricultural and Rural Cooperation ACP-EU (CPA). Postbus 380, 6700 AJ Wageningn, The Nethrlands.

Ofori, C.S., Nanday, S. (1969). The Effect of Method of Soil Cultivation on Yield and fertiliser response of maize grown on a forest ochrosol. Ghana J. Agric. Sci. 2: 19-24

Okwach, G. E., Simiyu, C. S. (1999). Effects of Land Management on runoff, Erosion and Crop Production in a Semi-arid Area of Kenya. E. Afr. Agric. For. J. 65(2), 125-142

Omer, A., Elamin, M., 1997. Effect of tillage and contour diking on sorghum establishment and yield on sandy clay soil in Sudan. Soil Tillage Res. 43 (3-4), 231-242.

Osunbitan, J.A., Oyedele, D.J., Adekalu, K.O., 2005. Tillage effects on bulk density, hydraulic conductivity and strength of a loamy sand soil in southwestern Nigeria. Soil and Till. Res.82, 57-64.

Ozpinar, S., Cay, A. 2006. Effect of different tillage systems on the quality and crop productivity of a clay–loam soil in semi-arid north-western Turkey. Soil and Till. Res. 88, 95–106.

Papendick, R.I., McCool, D.K., 1994. Residue management strategies—Pacific Northwest. In: Hartfield, J.L., Stewart, B.A. (Eds.), Crop Residue Management. Lewis Publishers, Boca Raton, FL, pp. 1–14.

Pathak, B.S. (1987). Survey of agricultural implements and crop production techniques. FAO Field Document 2, Eth/82/004. EARO, P. O. Box 2003, Addis Ababa, Ethiopia. 36p.

Penman, H.L. 1953. The physical basis of irrigation control. In: Report of the 13th international horticultural congress 1952. Synge, P.M. (ed.). Roy. Horticultural Soc., London, vol:II. pp 913-924.

Perillo, C.A., Gupta, S.C. Nater, E.A. Moncrief, J.F. 1999. Prevalence and initiation of preferential flow paths in a sandy loam with argillic horizon. Geoderma 89, 307–331.

Petersena, C.T., Jensena, H.E., Hansena, S. Koch, C. B. 2001. Susceptibility of a sandy loam soil to preferential flow as affected by tillage. Soil and Tillage Research 58, 81-89.

Rao, K.P.C., Steenhuis, T.S., Cogle, A.L., Srinivasan, S.T., Yule, D.F., Smith, G.D. 1998. Rainfall infiltration and runoff from an Alfisol in semi-arid tropical India. I. No-till systems. Soil and Till. Res. 48, 51-59.

Reicosky, D. C. (2001). Tillage-induced CO_2 Emissions and Carbon Sequestration: Effect of Secondary Tillage and Compaction. In: 1 World Congress on Conservation Agriculture, Madrid, 1-5 October, 2001. XUL Avda. Medina Azahara 49, pasaje, 14005 Cordoba, Spain. Pp265-274.

Richards, L.A., 1931: Capillary conduction of liquids in porous mediums. Physics. 1:318-333.

Ritsema, C. J., Dekker, L.W. 2000. Preferential flow in water repellent sandy soils: principles and modeling implications. Journal of Hydrology 231–232, 308–319

Rockström, J. 2003. Water for food and nature in drought-prone tropics: vapour shift in rain-fed agriculture. *Phil. Trans. R. Soc. Lond.* B 358, 1997–2009.

Rockström, J., Jansson, P-E., Barron, J., 1998. Seasonal rainfall partitioning under runon and surface runoff conditions on sandy soil in Niger. On-farm measurements and water balance modelling. J. Hydrol., 210: 68-92.

Rockström, J., Jansson, L. (1999). Conservation Tillage Systems for Dryland Farming: On-farm Research and Extension Experiences. E. Afr. Agric. For. J. 65(2), 101-114.

Rockström, J., Kaumbutho, P., Mwalley, P., Temesgen, M. (2001). Conservation tillage Farming Among Small-holder Farmers in E. Africa: Adapting and adopting innovative Land Management Options. In: World Congress on Conservation Agriculture, Madrid, 1-5 October, 2001. XUL Avda. Medina Azahara 49, pasaje, 14005 Cordoba, Spain. Pp 363-374.

Rockström, J. 1997. On-farm agrohydrological analysis of the Sahelian yield crisis: rainfall partitioning, soil nutrients and water use efficiency of pearl millet. PhD thesis. Natural Resources management department of systems Ecology. Stockholm University. Stockholm, Sweden.

Rockström, J., Valentin, C. (1997). Hillslope dynamics of on-farm water flows: The case of rain-fed cultivation of pearl millet on sandy soil in the Sahel. Agricultural Water Management. 33, 183-210.

Roth, C.H., Meyer, B., Frede, H.-G. and Derpsch, R., 1988. Effect of mulch rates and tillage systems on infiltrability and other soil physical properties of an Oxisol in Paran~, Brazil. Soil Tillage Res., 11: 81-91.

Rowland, J.R.J (ed). (1993). Dryland Farming in Africa. Published by Macmillan Education Ltd. in cooperation with the CTA, Postbus 380, 6700 AJ Wageningen, The Netherlands. 83p.

Russel, J.C., 1939. The effect of surface cover on soil moisture losses by evaporation. Soil Sci. Soc. Am. Proc. 4, 65–70.

Sall S., D. Norman and A.M. Freatherstone., 2000. Qualitative assessment of improved rice variety adoption: farmers' perspective. Agricultural Systems 66, pp 129-144.

SAS Institute Inc., 1999. SAS/STAT User's Guide, Version 8. Cary, NC, USA.

Savenije, H.G. 1997. Determination of evaporation from a catchment water balance at a momthly time scale. Hydrology and Earth System Sciences. 1, 93-100.

Savenije, H.G. 2004. The importance of interception and why we should delete the term evapotranspiration from our vocabulary. Hydrol. Process. 18, 1507-1511.

Scopel, E., Findeling, A. (2001). Conservation Tillage Impact on Rain fed maize in Semi-arid Zones of Western Mexico: Importance of runoff Reduction. In: 1 World Congress on Conservation Agriculture, Madrid, 1-5 October, 2001. XUL Avda. Medina Azahara 49, pasaje, 14005 Cordoba, Spain. Pp673-682

Scopel, E., Tardieu, F., Ema, G.O., Sebillotte, M. (2001). Effects of conservation tillage on water supply and rain fed maize production in semi arid zones of West-Central Mexico. NRG paper 01-01. Mexico, D.F.: CMMYT. pp32-47.

Smith, G.D., Coughlan, K.J., Yule, D.F., Laryea, K.B., Srivastava, K.L., Thomas, N.P., Cogle, A.L. 1992. Soil management options to reduce runoff and erosion on a hardsetting Alfisol in the semi-arid tropics. Soil and Till. Res. 25:195-215.

Smith, G.D., Coughlan, K.J., Yule, D.F., Laryea, K.B., Srivastava, K.L., Thomas, N.P. and Cogle, A.L., 1992. Soil management options to reduce runoff and erosion on a hardsetting Alfisol in the semi-arid tropics. Soil Tillage Res., 25:195-215.

Sojka, R.E., Westermann, D.T., Brown, M.J. and Meek, B.D., 1993. Zone-subsoiling effects on infiltration, runoff, erosion, and yields of furrow-irrigated potatoes. *Soil Tillage Res.,* 25: 351-368.

Souchere, V., King, D., Daroussin, J., Papy, F., Capillon, A. 1998. Effects of tillage on runoff directions: consequences on runoff contributing area within agricultural catchments. Journal of Hydrology 206, 256-267.

Steiner, K.G. (ed). (1998). Conserving Natural resources and Enhancing Food Security by Adopting No-Tillage. An Assessment of the Potential for Soil-conserving Production Systems in Various Agro-ecological Zones of Africa. Published by GTZ. Postfach 5180. D-65726 Eschborn. 47p.

Stewart, D. W., Dwyer, L. M. 1999. Mathematical Characterization of Leaf Shape and Area of Maize Hybrids. Crop Sci. 39:422–427.

Su, Y., Zhao, H., Zhang, T., Zhao, X. 2004. Soil properties following cultivation and non-grazing of a semi-arid sandy grassland in northern China. Soil and Till. Res. 75, 27–36.

Taa, A., Tanner, D. G., Gorfu, A. (1992). The Effects of Tillage Practice on Bread Wheat in three Different Cropping Sequences in Ethiopia. In: Tanner, D. G. (ed). 1992. Proceedings of the Seventh Regional Wheat Workshop for Eastern, Central and Southern Africa. Addis Ababa, Ethiopia: CIMMYT. pp 376-386.

Taddele, Z., 1994. *tef* in the Farming Systems of the Ada Area. Research Report No. 24. Institute of Agricultural Research, Addis Ababa, Ethiopia.

Tadele, Z., Aregu, L., Adela, A. (1999). Effect of Tillage Systems on *tef* Prodcuction. In: Progress Report of Holleta Research Center for the period April 1997 to March 1998. Ethiopian Agricultural Research Organization (EARO). Addis Ababa, Ethiopia. Pp 178-180.

Tadesse, N., Ahmed, S., Hulluka, M. (1994). The Effect of Minimum Tillage on Weed Management and Yield of Durum Wheat in Central Ethiopia. In: Tanner, D. G. (ed). 1994. The Eighth Regional Wheat Workshop for Eastern, Central and Southern Africa. Addis Ababa, Ethiopia: CIMMYT. Pp241-246

Tarawali, S.A., Peters, M., Mohammed-Salem, M.A. (1989). Improving Pasture Resources in West Africa: Evaluation of Forage Legumes to meet the Social and Ecological Constraints. In: Proceedings of the XVI International Grassland Congress, Nice, France. Pp 1495-1496.

Tarekegne, A., Gebre, A., Tanner, D. G., Mandefro, C. (1996). Effect of Tillage Systems and Fertilizer Levels on Continous Wheat Production in Central Ethiopia. In: Tanner, D. G., Payne, T. S., and Abdalla, O. S. (eds). 1996. The Ninth Regional Wheat Workshop for Eastern, Central and Southern Africa. Addis Ababa, Ethiopia: CIMMYT. Pp56-63

Teklu, E. and Gezahegn, A. (2003): Indigenous Knowledge and Practices for Soil and Water Management in East Wollega, Ethiopia. In: Wollny, Brodbeck, F.,

Howe, I. (eds): Technological and Institutional Innovations for Sustainable Rural Development. Deutscher Tropentag 2003. Göttingen, October 8-10, 2003 www.tropentag.de.

Temesgen, M. (1995). Assessing Field Performance of Animal-Drawn Plows. Technical Manual No. 9. Institute of Agricultural Research, Addis Ababa, Ethiopia. 29p

Temesgen, M. (2000). Animal Drawn Implements for Improved Cultivation in Ethiopia: Participatory Development and Testing. In: Kaumbutho, P.G., Pearson, R.A and Simalenga, T.E. (editors), 2000. *Empowering Farmers With Animal Traction.* Proceedings of the Workshop of the Animal Traction Network for Eastern and Southern Africa (ATNESA) held 20-24 September 1999, Mpumalanga, South Africa. Pp70-75.

Temesgen, M. (2001). Farmer Participatory Research Experiences: Melkassa Research Center. In: Ejigu Gonfa and Barry Pound (eds). 2001. Institutionalization of Farmer Participatory Research in The Southern Nations and Nationalities Peoples Regional State. Proceedings of the Second Forum on Farmer Participatory Research, June 29 – July 1, 2000. Awasa, Ethiopia. Pp13-15

Tiscareno, L.M., Gonzalez, B.A.D., Velazquez, V.M., Potter, K.N. (1999). Agricultural Research for Watershed Restoration in Central Mexico. Journal of Soil and Water Conservation. Ankeny. 54:4, 686-692

Tsai, Y.F., Huang, S.C., Lay, W.L. (1989). Effects of green manure on the growth of spring sorghum. Bulletin of the Taichung District Agricultural Improvement Station 23: 11-20

Van Reeuwijk, L.P., 1993. Procedures for Soil Analysis. International Soil Reference and Information Center, Netherlands.

Vogel, H., 1993. Tillage effects on maize yield, rooting depth and soil water content on sandy soils in Zimbabwe. *Field Crops Res.,* 33: 367-384.

Wanders, A.A., Stevens, P.A., 2000. Technology transfer and on-farm evaluation of animal powered equipment: approach and experience of IMAG-DLO. In: Kaumbutho P G, Pearson R A and Simalenga T E (eds), 2000. Empowering Farmers with Animal Traction. Proceedings of the workshop of the Animal Traction Network for Eastern and Southern Africa. (ATNESA) held 20-24 September 1999, Mpumalanga, South Africa.

West, T.O., Post, W.M., 2002. Soil organic carbon sequestration rates by tillage and crop rotation: a global data analysis. Soil Sci. Soc. Am. J. 66, 1930–1946.

Whiteman, P.T.S. (1979). Mekele Research Station, Northern Region: Progress Report April 1975-December 1976. Institute of Agricultural Research. P.O.Box 2003, Addis Ababa. Pp8-12

Willcocks, T.J. (1984). Tillage Requirements in Relation to Soil Type in Semi-Arid Rainfed Agriculture. Journal of Agricultural Engineering Research 30, 327-336.

Wilson, G.F., Lal, R., Okigo, B.N. (1982). Effects of cover crops on soil structure and yield of subsequent arable crops grown under strip tillage in an arable eroded Alfisol. Soil Till. Res. 2:233-250

Zekaria, S., 2002, Innovative and Successful Technical Experience in the Production of Agricultural Statistics and Food Security of Ethiopia, p.16, Central Statistical Authority, Addis Ababa.

Acknowledgments

Dr. Johan Rockstrom has guided me on conservation tillage research work since 1998. His dream to solve the problems of water productivity among smallholder farmers in sub-Saharan Africa, in general, and in Ethiopia, in particular, is beyond imagination. He has always spared time for the research despite so much workload. He has been with me in the field laying out the experimental set ups before and after I started my PhD studies. We have had long discussions and heated debates at times that all led to a fruitful result we see today. I am greatly indebted to him.

My great appreciation goes to my promoter Prof. dr. ir. H.H.G. Savenije. Since we met at UNESCO-IHE in 2002 he has been inspired by the on-farm research I have been conducting with farmers. He has guided me superbly on how to conduct scientific research and how to write articles. His positive and yet detailed critical assessment of my work and my writings is what has resulted in the present shape of this discourse. I consider him a real friend of smallholder resource poor farmers and researchers of the same environment. Prof. Savenije was also my mentor after Mr, Schotanus left UNESCO-IHE.

Dr. Willem Hoogmoed supervised my research work and writings from the beginning of my PhD studies. He gave me so many useful hints on how to conduct research in the field. He has been very quick in responding to my technical questions and spent his time with me with great hospitality whenever I visited him at Wageningen University. He collected so many research articles from reputed journals for me to review. His critical comments led to significant improvements to both the research work and manuscript preparations.

I am also indebted to my mentor Mr. Daniel Schotanus who facilitated the administrative and organizational issues at UNESCO-IHE until July 2006. My special thanks go to Ms. Jolanda Boots who carried out the rather complicated administrative and financial arrangements perfectly. I also thank Ms. Vandana Sharma who wonderfully administered my scholarship at the beginning of my studies. I would also like to thank Mrs Teresa Ogenstad, Mrs Gunnel Olofsson and Mr. Goran Axberg who helped me a lot when I stayed at Stockholm Environment Institute during May-June, 2006. Mrs. Louise Karlberg helped me in setting up and running the CoupMmodel. I appreciate her time and devotion under difficult situations. My appreciation also goes to Miriam Gerrits who helped me a lot in the layout and styling of the thesis. Dr. Belay Semmanni, Dr. Kidane Georgis, Dr. Sahle Medhin Sertsu and Dr. Taye Bekele gave me valuable comments on my research proposal.

The field research would not have been carried out had it not been for the assistance I got from farmers Hawas Gelete, Nesru Wajo, Kumbi Wari, Mengiste Wale, Zilelew Mengiste, Sisay Tekle Sillase, Tamiru Tekle Sellase and Shambel Waqo who all contributed to the research work. Befrdu Taddele, Zewdu Abebe and Nigussie T/Michael assisted in data collection.

My special thanks are extended to staff of Melkassa Research Center specially Dereje Mersha and Senbaba Gutema who helped me in Laboratory analysis and Tewodros Mesfin who also helped me in laboratory works and in organizing long term meteorological data.

I am also indebted to Dr. Abayneh and Mr. Abebe of the National Soil Laboratory who carried out laboratory analysis on the physical and chemical properties of soil samples. My thanks are also extended to Mr. Yohannes of the metrology department of the Ethiopian Agricultural Research Institute for analyzing my data using SAS.

The study was financed by the Dutch Foundation for the Advancement of Tropical Research (WOTRO), the Regional Land Management Unit (RELMA) of SIDA and the Ethiopian Agricultural Research Institute (EARI).

Samenvatting

Traditioneel wordt er in Ethiopië met de *Maresha* ploeg geploegd. Deze ploeg vereist veelvuldig en kruislings ploegen. Deze methode is niet efficiënt voor het maximaal benutten van regenval. In de semi-aride gebieden van Afrika is het cruciaal dat maximaal gebruik wordt gemaakt van de beschikbare regenval.

Er zijn een aantal innovaties ontwikkeld om de effectiviteit van het ploegen te verhogen. Omdat de boeren om allerlei redenen gebruik willen blijven maken van de *Maresha* ploeg, zijn deze innovaties in dit onderzoek aangebracht op de traditionele ploeg. Deze innovaties zijn uitgetest op proefvelden waar Mais en het traditionele *Tef* gewas worden verbouwd. Van de proefvelden is een volledige waterbalans gemaakt en is de productie nauwkeurig gemonitord.

De tests hebben laten zien dat met de aangepaste ploeg niet alleen de landbouwproductiviteit toenam, maar dat ook de hoeveelheid energie benodigd voor het ploegen minder was, en dat het percentage regenwater dat de plant ten goede kwam, toenam. Door het onderzoek samen met de boeren uit te voeren is bovendien bewerkstelligd dat de nieuwe technologie van harte door de gemeenschap is overgenomen.

About the author

Melesse Temesgen was born on 7 July 1964 in Gojjam, Ethiopia. He completed his primary and secondary school in 1980. In the same year, he joined the then Alemaya College of Agriculture, Addis Ababa University, and graduated in July 1984 with distinction in Agricultural Engineering. He joined the then Institute of Agricultural Research in September 1984. He started his post graduate studies in 1985 and got his MSc degree, in Agricultural Engineering, in September 1987 from the University of Newcastle upon Tyne (United Kingdom). He returned to Ethiopia and joined the same institution. He has worked in different positions as a researcher for 15 years before starting his PhD studies. Between 1997 and 2001 he was given an additional task of co-coordinating the National Agricultural Mechanization Research Program.

He has presented research papers in several national and international workshops and congresses and has published research articles in technical manuals, proceedings and international journals. He has developed about 10 types of successfully tested agricultural implements most of which are used for land preparation with the objective of improving rain water management such as soil moisture conservation through reduction of surface runoff and evaporation.

On June 30, 2002, he received the first national prize for outstanding achievement in science and technology in agriculture from the president of the federal government after winning an open competition organized by the Ethiopian Science and Technology Commission. He started his PhD studies in 2002 at UNESCO-IHE and at the department of Water Resources in TU Delft, The Netherlands. He got the scholarship from The Netherlands Fund for Tropical Agriculture (WOTRO). His field research was financed by the Regional Land Management Unit (RELMA) of Sida. He hopes to share his knowledge and experience on conservation tillage with smallholder farmers in the drought prone semi-arid Ethiopia.